STUDY GUIDE
VOLUME ONE: CHAPTERS 1–16

SEARS & ZEMANSKY'S

COLLEGE PHYSICS

8TH EDITION

YOUNG & GELLER

Laird Kramer
Florida International University

PEARSON

Addison
Wesley

San Francisco   Boston   New York
Cape Town   Hong Kong   London   Madrid   Mexico City
Montreal   Munich   Paris   Singapore   Sydney   Tokyo   Toronto

Editor-in-Chief:	Adam Black
Associate Editor:	Chandrika Madhavan
Managing Editor, Production:	Corinne Benson
Production Supervisor:	Jane Brundage
Executive Marketing Manager:	Christy Lawrence
Manufacturing Buyer:	Pam Augspurger
Production Services:	WestWords, Inc.
Cover Design:	Seventeenth Street Studios

Cover Photo Credit: © Ken Findlay/Dorling Kindersley

ISBN 0-8053-9222-X

1 2 3 4 5 6 7 8—BRG—10 09 08 07 06

www.aw-bc.com

CONTENTS

Preface v

Chapter 1   Models, Measurements, and Vectors 1
Chapter 2   Motion Along a Straight Line 12
Chapter 3   Motion in a Plane 21
Chapter 4   Newton's Laws of Motion 31
Chapter 5   Applications of Newton's Laws 42
Chapter 6   Circular Motion and Gravitation 58
Chapter 7   Work and Energy 66
Chapter 8   Momentum 77
Chapter 9   Rotational Motion 86
Chapter 10   Dynamics of Rotational Motion 96
Chapter 11   Elasticity and Periodic Motion 107
Chapter 12   Mechanical Waves and Sound 115
Chapter 13   Fluid Mechanics 122
Chapter 14   Temperature and Heat 129
Chapter 15   Thermal Properties of Matter 136
Chapter 16   The Second Law of Thermodynamics 145

PREFACE

What does an Olympic athlete, your favorite music artist, and Albert Einstein have in common? They all became experts in their fields through practice. To understand physics and to do well in your course, you must practice. When you learned to walk, ride a bike, and drive a car; you had to practice to master those skills. It would be silly to think you can learn physics by listening to lectures and skimming the book. This study guide is designed to help you practice and to build a deep understanding of physics.

Expert problem solvers in physics follow a systematic approach in their problem solving. Elite athletes also follow a systematic approach in their training to reach the upper level of their sport. You should also follow a systematic approach in your physics course to fully develop your skills. To encourage you in building good problem-solving skills, this study guide follows a systematic problem-solving procedure throughout—the **Set Up, Solve,** and **Reflect** procedure developed in the textbook.

In the **Set Up** phase of the problem, you should examine what you know, what you need to find, and plan a strategy for tackling the problem. This step often benefits greatly from sketches that will guide you through the problem. Rarely do physicists discuss cutting-edge research problems without first sketching their ideas. You should also indicate any assumptions you'll make during the solution at this step.

When you proceed to **Solve** the problem, start with your procedure, and identify what physics principles you will be using. Then work through the solution step-by-step. Write down all of your work so you may return and check it later. If you run into a dead end, don't erase your work as you may find it useful in a later phase of the problem. Try another avenue when you get stuck and you will eventually find the solution.

After completing the problem, **Reflect** and check over your work. You'll want to make sure the answers make sense—if you're estimating how high an elephant can jump, you'd expect it ought to be less than a meter or two. This also gives you a chance to examine what you have learned from the problem. Consider how this problem compares to the last problem you completed, the example in the text, and the example shown in class. Much can be learned in physics by comparing a variety of problems and noting their similarities and differences.

Questions and problems chosen for this study guide cover the most critical topics you'll encounter. Working through the guide will better prepare you for homework (including MasteringPhysics) and exams and assist in developing a deeper understanding of physics. One way to build confidence is to try working through the questions and problems in the guide for practice, referring to the solutions only when you get stuck. Building confidence before an exam reduces stress during the exam, improving performance. Summaries, Objectives, Concepts and Equations, and Problem Summaries are ancillary materials that help bring the physics topics of each chapter into coherence. Taking advantage of all of the components in this study guide will help build your problem-solving repertoire.

This study guide is but one of many resources at your disposal when learning physics. Your instructor, class, and textbook are also important resources. But you should also consider who approaches the material from the same level—your fellow students. The best untapped resource is often other students learning physics for the first time. Discuss physics as a group and confront your questions together, just as many professionals collaborate in the workplace.

We know physics has a reputation for being challenging. While it can be challenging, many students have succeeded in learning physics. Their success was built on a series of small steps, regular practice, and following a systematic approach. Follow their footsteps and you will master physics as they did. You'll also find physics to be a rich and beautiful subject.

Good luck and enjoy learning physics!

Laird Kramer
Miami, Florida, 2006

1

MODELS, MEASUREMENTS, AND VECTORS

Summary

Physics is the study of natural phenomena. In physics, we build theories through observing nature, and those theories evolve into physical laws. We often seek simplicity; models are simplified versions of physical phenomena that allow us to gain insight into a physical process. This chapter begins with the foundations of measurement, including the International System (SI), units, conversions, precision, significant figures, estimates, orders of magnitude, and scientific notation. We will also examine physical quantities. Scalar quantities, such as temperature, are described by a single number. Vector quantities, such as velocity, are described by both a magnitude and a direction. Finally, we review how to combine vector quantities by breaking them into their components. Techniques we develop in this chapter will be used throughout our investigation of physics.

Objectives

- Understand the process of experimentation, theory, and laws.
- Know the SI units for length, mass, and time, and know common metric prefixes.
- Know how to express results with proper units and how to convert between different sets of units.
- Understand uncertainties and how significant figures express precision.
- Understand scalar and vector quantities.
- Know how to add and subtract vectors by the graphical method.
- Know how to add and subtract vectors by the component method.

Concepts and Equations

Term	Description
Physical Law	A physical law is a well-established physical principle.
Model	A model is a simplified version of a physical system that is used to concentrate on the most important features of the system.
Système International (SI)	The Système International (SI) is the system of units based on the metric system. It established refined definitions of units, including the second, meter, and kilogram.
Significant Figures	The number of significant figures is the number of meaningful digits in a number. In multiplication or division, the number of significant figures in the result is no greater than in the factor with the fewest significant figures. In addition or subtraction, the result can have no more decimal places than the term with the least number of decimal places.
Scalar Quantity	A scalar quantity is expressed by a single number. Examples include temperature, mass, and time.
Vector Quantity	A vector quantity is expressed by both a magnitude and a direction and is shown as an arrow in sketches. Vectors are often represented as single letters with arrows above them or in boldface type. Common examples include velocity, displacement, and force.
Trigonometric Functions	The basic trigonometric functions relate the lengths of the sides of a right triangle to the inside angle. We define $\sin\theta$, $\cos\theta$, and $\tan\theta$ as follows for the right triangle shown: $$\sin\theta = \frac{\text{opposite side}}{\text{hypotenuse}}, \quad \cos\theta = \frac{\text{adjacent side}}{\text{hypotenuse}}, \quad \tan\theta = \frac{\text{opposite side}}{\text{adjacent side}}.$$ We can also write, for any angle θ, $$\tan\theta = \frac{\sin\theta}{\cos\theta} \text{ and } \sin^2\theta + \cos^2\theta = 1.$$
Component Vectors	The vector $\vec{A}$ lying in the x-y plane can be represented as the sum of the vector $\vec{A}_x$ parallel to the x axis and the vector $\vec{A}_y$ parallel to the y axis; $\vec{A}_x$ and $\vec{A}_y$ are the x and y component vectors of $\vec{A}$.
Magnitude of a Vector	The magnitude of a vector is the length of the vector. Magnitude is a scalar quantity that is always positive. It has several representations, including $$\text{Magnitude of } \vec{A} = A = |\vec{A}|.$$

The magnitude can be found from the component vectors:

$$A = \sqrt{A_x^2 + A_y^2}.$$

| **Vector Addition and Subtraction** | Vectors are added graphically by placing the tail of the second vector at the head of the first vector. |
| | |

Vectors are subtracted by reversing the direction of the vector to be subtracted and then adding.

Vectors can also be added by using component vectors. For components A_x and A_y of the vector $\vec{A}$ and components B_x and B_y of the vector $\vec{B}$, the components R_x and R_y of the resultant vector $\vec{R}$ are given by

$$R_x = A_x + B_x \text{ and } R_y = A_y + B_y.$$

Conceptual Questions

1: Sketch the situation

A man uses a cable to drag a trunk up the loading ramp of a mover's truck. The ramp has a slope angle of 20.0°, and the cable makes an angle of 30.0° with the ramp. Make a sketch of this situation.

Solution

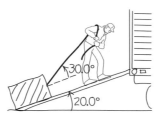

▲ **Figure 1.1** Sketch of trunk being dragged up a loading ramp.

SET UP AND SOLVE The sketch is shown in Fig 1.1. The ramp makes an angle of 20° with the ground. The trunk is on the ramp and the cable is attached to the trunk. The cable makes an angle of 30° with respect to the ramp, clearly marked. The mover is shown pulling the trunk up the ramp.

REFLECT Understanding the physical situation in physics problems is critical for a correct interpretation. You should always draw a diagram (or diagrams) of the physical system you are investigating. Even when a figure is provided, it is often useful to sketch the important aspects. Only after creating a diagram should you proceed to interpret the physics and determine the proper equations to apply.

2: Dimensional analysis practice

Based *only* on consistency of units, which of the following formulas could *not* be correct? In each case, x is distance, v is speed, and t is time.

(a) $t = \sqrt{\dfrac{2x}{9.8 \text{ m/s}^2}}$

(b) $x = vt + (4.9 \text{ m/s}^2)t$

(c) $v = v_0 \sin\theta + \dfrac{(9.8 \text{ m/s}^2)x}{v_0 \cos\theta}$

(d) $x^2 - \dfrac{2v_0^2 \sin\theta}{9.8 \text{ m/s}^2} - \dfrac{2v^2}{9.8 \text{ m/s}^2} = 0$

(e) $v^2 = v_0^2 - 2(9.8 \text{ m/s}^2)(v_0 \tan\theta - \frac{1}{2}(9.8 \text{ m/s}^2)t^2)$

(f) $t = \dfrac{v^2 + (4.9 \text{ m/s}^2)x}{(3.0 \text{ m/s}^2)x}$

Solution

SET UP AND SOLVE For each of the six equations, carefully examine the units of each term in the equation. Equations (*a*), (*c*), and (*d*) are correct dimensionally; however, equations (*b*), (*e*), and (*f*) are incorrect. The rightmost term in equation (*b*) has units of (m/s), while the other two terms have units of (m). In equation (*e*), the leftmost term inside the parentheses on the right ($v_0 \tan\theta$) has units of (m/s), which, when combined with the (m/s^2) outside of the left parenthesis, would result in units of (m^2/s^3). The other three terms in the equation have units of (m^2/s^2). The fraction in equation (*f*) has no units, while, on the left-hand side of the equation, has units of (s).

Thus, an error exists in each of the three equations (*b*), (*e*), and (*f*), since the units on the two sides of the equation do not agree. The next step would be to recheck our derivation to locate the source of the mistake.

REFLECT Dimensional analysis is a powerful technique to help keep you from making errors. Catching the three errors in this problem would save you time and credit while reducing confusion. Always check your units!

3: Maximum and minimum magnitudes of a vector

Given vector $\vec{A}$ with magnitude 1.3 N and vector $\vec{B}$ with magnitude 3.4 N, what are the minimum and maximum magnitudes of $\vec{A} + \vec{B}$?

Solution

SET UP AND SOLVE The maximum magnitude occurs when the two vectors are parallel and point in the same direction. The minimum magnitude occurs when the two vectors are parallel and point in the opposite direction (called antiparallel).

For parallel vectors, the magnitude is the sum of their magnitudes, 4.7 N in this case. For antiparallel vectors, the magnitude is the difference in their magnitudes, 2.1 N here.

REFLECT This example helps illustrate the fact that vectors do not add like ordinary scalar numbers. They do not subtract like scalar numbers either. The magnitude of $\vec{A} + \vec{B}$ for any arbitrary alignment of the two vectors must lie between 2.1 N and 4.7 N.

Problems

1: Convert knots to m/s

A yacht is traveling at 18.0 knots. (One knot is 1 nautical mile per hour.) Find the speed of the yacht in m/s.

Solution

SET UP We'll use a series of conversion factors to solve this problem. Appendix E gives 1 nautical mile = 6080 ft and 1 mi = 5280 ft = 1.609 km. We know that 1 km = 1000 m and that 1 hour = 60 min = 60 × (60 s) = 3,600 s.

SOLVE We apply the conversion factors to the speed in knots to solve:

$$18.0 \text{ knots} = \left(\frac{18.0 \text{ nautical miles}}{1 \text{ h}}\right)\left(\frac{6{,}080 \text{ ft}}{1 \text{ nautical mile}}\right)\left(\frac{1.609 \text{ km}}{5{,}280 \text{ ft}}\right)\left(\frac{1{,}000 \text{ m}}{1 \text{ km}}\right)\left(\frac{1 \text{ h}}{3{,}600 \text{ s}}\right) = 9.26 \text{ m/s}.$$

REFLECT Using a combination of several conversion factors, we have found that 18.0 knots is equivalent to 9.26 m/s. Each of the quantities in parentheses is equal to unity; hence, multiplying 18.0 knots by several factors of unity doesn't change the physical meaning of the quantity. Crossing out the units helps prevent mistakes.

2: Finding components of vectors

Find the x and y components of the vector in Figure 1.2.

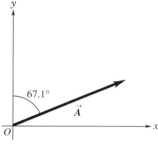

▲ **Figure 1.2** Problem 2.

Solution

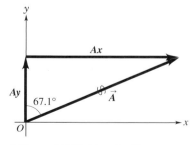

▲ **Figure 1.3** Problem 2 with components.

SET UP We find the components of a vector by examining the triangle made by the vector and the coordinate axes. Figure 1.3 shows Figure 1.2 redrawn to include the component vectors.

SOLVE The x component of $\vec{A}$ is located opposite the 67.1° angle; hence, we'll use the sine relation:

$$A_x = A \sin 67.1° = (26.2 \text{ cm}) \sin 67.1° = 24.1 \text{ cm.}$$

The y component of $\vec{A}$ is located adjacent to the 67.1° angle; thus, we'll use the cosine relation:

$$A_y = A \cos 67.1° = (26.2 \text{ cm}) \cos 67.1° = 10.2 \text{ cm.}$$

The vector has an x component of 24.1 cm and a y component of 10.2 cm.

REFLECT Finding the components of the vector required applying the sine and cosine relations. Often, but not always, the horizontal components will use the cosine relation and the vertical components will use the sine relation. This example is an exception to that general assertion. It is important to examine a problem carefully to identify the proper relation for each component.

3: Vector addition

Find the vector sum $\vec{A} + \vec{B}$ for the two vectors in Figure 1.4. Express the results in terms of components.

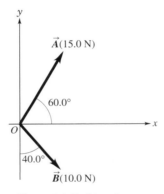

▲ **Figure 1.4** Problem 3.

Solution

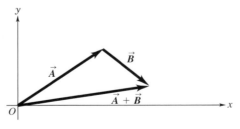

▲ **Figure 1.5** Sketch of Problem 3.

SET UP Figure 1.5 shows a sketch of the two vectors added together, head to tail. The sketch shows that we should expect a resultant in the first quadrant, with positive x and y components. We will add the vectors by adding their x and y components, using the Cartesian coordinate system provided.

SOLVE We find the components of the vectors by examining the triangles made by the vectors and their components. For $\vec{A}$,

$$A_x = A\cos 60.0° = (15.0\text{ N})\cos 60.0° = 7.50\text{ N},$$
$$A_y = A\sin 60.0° = (15.0\text{ N})\sin 60.0° = 13.0\text{ N}.$$

For $\vec{B}$,

$$B_x = B\sin 40.0° = (10.0\text{ N})\sin 40.0° = 6.43\text{ N},$$
$$B_y = -B\cos 40.0° = -(10.0\text{ N})\cos 40.0° = -7.66\text{ N}.$$

We can now sum the components:

$$R_x = A_x + B_x = 7.50\text{ N} + 6.43\text{ N} = 13.9\text{ N},$$
$$R_y = A_y + B_y = 13.0\text{ N} - 7.66\text{ N} = 5.34\text{ N}.$$

The resultant vector has an x component of 13.9 N and a y component of 5.34 N.

REFLECT The resultant vector has positive components and resides in the first quadrant, as expected. Note how the components of the two vectors included both sine and cosine terms; that is, vector A's x component included the cosine component and vector B's x component included the sine component. This results from how the vectors' angles were given: Vector A's angle was with respect to the horizontal axis and vector B's angle was with respect to the vertical axis. It is critical not to automatically associate all horizontal components with the cosine and all vertical components with the sine.

Practice Problem: Find the magnitude and direction of the resultant vector. *Answer:* the magnitude is 14.9 N, and its direction is 21.0° above the positive x axis.

4: Determine displacement on a lake

Marie paddles her canoe around a lake. She first paddles 0.75 km to the east, then paddles 0.50 km 30° north of east, and finally paddles 1.0 km 50° north of west. Find the resulting displacement from her origin.

Solution

SET UP Displacement is a vector indicating change in position. The displacement vector points from the starting point to the endpoint. If we represent each of the three segments of the journey as a vector, the displacement vector is the sum of the three vectors.

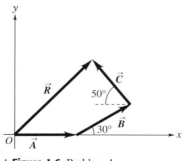

▲ **Figure 1.6** Problem 4.

Figure 1.6 shows a sketch of the three segments (labeled $\vec{A}$, $\vec{B}$, and $\vec{C}$) and the resultant displacement vector ($\vec{R}$). We will add the three vectors, using the Cartesian coordinate system in the figure.

SOLVE We find the components of the vectors by examining the triangles made by the vectors and their components. For $\vec{A}$, there is only a horizontal component:

$$A_x = A = 0.75 \text{ km},$$
$$A_y = 0.$$

For $\vec{B}$,

$$B_x = B\cos 30° = (0.50 \text{ km})\cos 30° = 0.443 \text{ km},$$
$$B_y = B\sin 30° = (0.50 \text{ km})\sin 30° = 0.250 \text{ km}.$$

And for $\vec{C}$,

$$C_x = -C\cos 50° = -(1.0 \text{ km})\cos 50° = -0.643 \text{ km},$$
$$C_y = C\sin 50° = (1.0 \text{ km})\sin 50° = 0.766 \text{ km}.$$

The x component is negative, as it points to the west. We can now sum the components:

$$R_x = A_x + B_x + C_x = 0.75 \text{ km} + 0.443 \text{ km} - 0.643 \text{ km} = 0.550 \text{ km},$$
$$R_y = A_y + B_y + C_y = 0 \text{ km} + 0.250 \text{ km} + 0.766 \text{ km} = 1.016 \text{ km}.$$

The resultant displacement vector has an x component of 0.55 km and a y component of 1.02 km. We can express the displacement vector in terms of magnitude and direction. To find the magnitude, we use the Pythagorean theorem:

$$R = \sqrt{R_x^2 + R_y^2} = \sqrt{(0.550 \text{ km})^2 + (1.016 \text{ km})^2} = 1.16 \text{ km}.$$

The inverse tangent gives us the angle:

$$\theta = \tan^{-1}\frac{R_y}{R_x} = \tan^{-1}\frac{1.016 \text{ km}}{0.550 \text{ km}} = 61.6°.$$

The resultant displacement vector has a magnitude of 1.16 km and points 61.6° above the positive x axis.

REFLECT Marie paddled a total of 2.25 km, only to end up 1.16 km away from her starting point. This problem shows how the magnitude of a vector sum can be smaller than the sum of the individual vector magnitudes. Note that we carried an extra significant figure through the calculations and rounded off only in the final step.

5: Review of simultaneous equations

We will sometimes encounter problems with two unknowns (x and y, for example). These types of problems require two equations to be solved together, or *simultaneously*. Typically, one equation is solved in terms of one variable and is then substituted into the other equation. In this case, solve one equation for x in terms of y, and then substitute the expression for x into the second equation to obtain an expression that involves only y.

For example, suppose we are given the following two equations:

$$27T_A + 13T_B = 0,$$
$$32T_A + 52T_B = 22.$$

Solve for T_A and T_B.

Solution

SET UP Both of the expressions involve two unknowns, and we cannot find a solution by using only one equation. We'll write T_B in terms of T_A in the first equation and then substitute for T_B in the second equation.

SOLVE Rewrite the first expression in terms of T_B:

$$T_B = -\frac{27T_A}{13}.$$

Substituting T_B into the second expression gives

$$32T_A + 52\left(-\frac{27T_A}{13}\right) = 22,$$

$$\left(32 + 52\left(-\frac{27}{13}\right)\right)T_A = 22,$$

$$-76T_A = 22,$$

$$T_A = -0.290.$$

Substituting the value for T_A back into either expression to find T_B yields

$$32(-0.290) + 52T_B = 22,$$

$$T_B = 0.601.$$

The two equations combine to give $T_A = -0.29$ and $T_B = 0.60$.

REFLECT An alternative solution is to multiply the first equation by 32, multiply the second by 27, and subtract the second equation from the first. This gives the same result, although requires more forethought. You may choose the method you prefer, and you may end up applying both to particular classes of problems.

If we encounter three unknowns in an expression, how many equations will we need to solve for each unknown simultaneously? Three equations will be needed to solve for the three unknown quantities.

6: Review of the quadratic formula

We will sometimes encounter problems in which the mathematical representation leads to a quadratic equation. In some cases, we'll need to use the quadratic formula to find the solution. For example, if a ball is tossed into the air, its position depends on its initial speed and the elapsed time. The ball's position can be given by

$$y = v_0 t - (4.9 \text{ m/s}^2)t^2,$$

where v_0 is the initial speed and t is the elapsed time. Imagine a ball tossed with an initial speed of 30.0 m/s. Find the elapsed time(s) when the ball is at a height of 12.5 m.

Solution

SET UP We recognize that the equation is quadratic, since it has a t^2 term, a t term, and a constant term. If we try to rewrite the equation in terms of t alone, we find that we cannot isolate the t term. We will have to employ the quadratic formula to solve the problem.

SOLVE We rewrite the equation, substituting the given values:

$$12.5 \text{ m} = (30.0 \text{ m/s})t - (4.9 \text{ m/s}^2)t^2.$$

The quadratic formula requires that the equation be written as $ax^2 + bx + c = 0$, so we rearrange the equation to obtain

$$-(4.9 \text{ m/s}^2)t^2 + (30.0 \text{ m/s})t - (12.5 \text{ m}) = 0.$$

From this rearrangement, we see that $a = 4.9 \text{ m/s}^2$, $b = 30.0 \text{ m/s}$, and $c = -12.5 \text{ m}$. The solutions of the quadratic equation are

$$x = \frac{-b \pm \sqrt{b^2 - 4ac}}{2a}.$$

Substituting our values into the quadratic equation gives

$$t = \frac{-(30.0 \text{ m/s}) \pm \sqrt{(30.0 \text{ m/s})^2 - 4(-4.9 \text{ m/s}^2)(-1.25 \text{ m})}}{2(-4.9 \text{ m/s}^2)}.$$

Multiplying out the terms and canceling the units produces

$$t = \frac{-(30.0 \text{ m/s}) \pm \sqrt{(900.0 - 245.0)(\text{m}^2/\text{s}^2)}}{-9.8(\text{m/s}^2)} = \frac{-30.0 \pm 25.59}{-9.8} \text{ s} = 0.445 \text{ s}, 5.67 \text{ s}.$$

There are two times when the ball is at a height of 12.5 m: 0.445 s and 5.67 s. These are, respectively, when the ball is rising to its maximum height and when it is falling from its maximum height.

REFLECT You must learn to recognize quadratic equations. Once you identify a quadratic equation, the solution is straightforward (although it requires careful algebra). Quadratic equations result in two solutions, and you must be able to interpret their meanings. In this case, the two solutions corresponded to the upward and downward motion of the ball. You may only need one of the solutions in a particular situation. If neither solution seems reasonable, then you should check your work. We'll first encounter quadratic equation problems in Chapter 2.

Problem Summary

The problems in this chapter give you a foundation that you will use throughout your physics course. Common elements in them will help you to develop good problem-solving techniques, including the following:

- Identifying a procedure for finding the solution
- Making a sketch when no figure is provided
- Adding appropriate coordinate systems to the sketch
- Identifying the known quantities in the problem
- Finding appropriate equations to solve for the unknown quantities
- Checking for unit consistency in derived equations
- Reflecting on the results to check for inconsistencies

We will see how these techniques apply to a wide variety of problems as we progress. While they may seem cumbersome right now, they will help you solve the problems you encounter.

2

MOTION ALONG A STRAIGHT LINE

Summary

We will introduce *kinematics,* the study of an object's motion, or change of position with time, in this chapter. Motion includes *displacement,* the change in position of an object; *velocity,* the rate of change of position of the object with respect to time; and *acceleration,* the rate of change of velocity of the object with respect to time. We introduce average velocity and acceleration as changes over a time interval and instantaneous velocity and acceleration as changes over an infinitely short time interval. We'll learn relationships among displacement, velocity, and acceleration and see how they are modified for freely falling objects. We'll restrict ourselves to motion along a straight line, or one-dimensional motion, in this chapter and expand our examination to motion in a plane in the next chapter. This is our first step into understanding mechanics, the study of the relationships among force, matter, and motion, that we'll cover in the upcoming chapters.

Objectives

- Learn the definitions of kinematic variables.
- Calculate average and instantaneous velocities.
- Calculate average and instantaneous accelerations.
- Learn and apply the equations of motion for constant acceleration
- Apply equations of motion for constant acceleration to freely falling objects.
- Find the velocities of objects relative to different reference frames.

Concepts and Equations

Term	Description
Displacement	An object's displacement is a vector quantity whose x component is $$\Delta x = x_2 - x_1,$$ where x_1 is the starting position and x_2 is the final position. The SI unit for displacements is meters (m).
Average Velocity	An object's average velocity is a vector quantity whose x component is defined as the x component of the displacement Δx, divided by the time interval Δt in which the displacement occurs: $$v_{av,x} = \frac{x_2 - x_1}{t_2 - t_1} = \frac{\Delta x}{\Delta t}.$$ The SI unit for velocity is meters per second (m/s).
Instantaneous Velocity	An object's instantaneous velocity is the limit of the average velocity as Δt goes to zero. The x component is defined as $$v_x = \lim_{t \to \infty} \frac{\Delta x}{\Delta t}.$$ The term *velocity* refers to the instantaneous velocity.
Average Acceleration	The average acceleration of an object as it moves from x_1 (at t_1) to x_2 (at t_2) is a vector quantity whose x component is the ratio of the change in the x component of velocity, $\Delta v_x = v_{2x} - v_{1x}$, to the time interval Δt: $$a_{av,x} = \frac{v_{2x} - v_{1x}}{t_2 - t_1} = \frac{\Delta v_x}{\Delta t}$$ The SI unit for acceleration is meters per second per second (m/s^2).
Instantaneous Acceleration	An object's instantaneous acceleration is the limit of the average acceleration as Δt goes to zero. The x component of acceleration is defined as $$a_x = \lim_{t \to \infty} \frac{\Delta v_x}{\Delta t}.$$ The term *acceleration* refers to the instantaneous acceleration.
Motion with Constant Acceleration	When an object moves with constant acceleration in a straight line along the x axis, the position, velocity, acceleration, and time are related by the following three equations: $$x = x_0 + v_{0x}t + \tfrac{1}{2}a_x t^2,$$ $$v_x = v_{0x} + a_x t,$$ $$v_x^2 = v_{0x}^2 + 2a_x(x - x_0).$$ Similar relations hold for motion along the y axis.
Freely Falling Object	A freely falling object is an object that moves under the influence of the gravitational force. The acceleration due to gravity is denoted by g, is directed downwards, and has a value of 9.8 m/s^2 near the surface of the earth.
Relative Velocity along a Straight Line	When an object P moves relative to a reference frame B, and B moves relative to a second reference frame A, the velocity of P relative to B is denoted by $v_{P/B}$, the velocity of P relative to A is denoted by $v_{P/A}$, and the velocity of B relative to A is denoted by $v_{B/A}$. These velocities are related by the formula $$v_{P/A} = v_{P/B} + v_{B/A}.$$

Conceptual Questions

1: Velocity and acceleration at the top of a ball's path

A ball is tossed vertically. (a) Describe the velocity and acceleration of the ball just before it reaches the top of its flight. (b) Describe the velocity and acceleration of the ball at the instant it reaches the top of its flight. (c) Describe the velocity and acceleration of the ball just after it reaches the top of its flight.

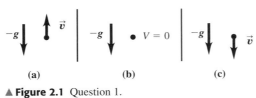

▲ **Figure 2.1** Question 1.

Solution

SET UP AND SOLVE Figure 2.1 shows the three time frames we will examine. During its flight, the ball undergoes acceleration due to gravity. The initial velocity is upwards, slowing to zero at the top of the flight and then increasing downwards.

Part (a): The velocity is upwards and very small just before the top of the flight. The acceleration due to gravity is downwards.

Part (b): The velocity is zero at the top of the flight. The acceleration due to gravity remains constant and downwards. The acceleration has caused the velocity to decrease from the small positive value in part (a) to zero.

Part (c): The velocity is downwards and very small just after the top of the flight. The acceleration due to gravity remains constant and downwards. The acceleration has caused the velocity to increase downwards from zero in part (b).

REFLECT The acceleration due to gravity causes a change in velocity during the flight. The ball starts with an upward velocity, which slows, drops to zero, and then increases downwards. The acceleration due to gravity is constant throughout the motion. The velocity is zero for an instant at the top, changing from slightly upward to slightly downward around this instant.

2: Comparing two cyclists

A graph of position versus time for two cyclists is shown in Figure 2.2. (a) Do the cyclists start from the same position? (b) Are there any times when the cyclists have the same velocity? (c) What is happening at the intersection of lines *A* and *B?*

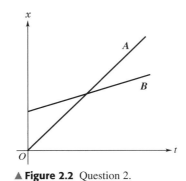

▲ **Figure 2.2** Question 2.

Solution

SET UP AND SOLVE **Part (a):** We find the starting location by examining the position when time is zero (i.e., by looking at the x intercept). At $t = 0$, the two cyclists have different locations.

Part (b): The velocity is found by examining the slope of the position-versus-time graph. The slopes of the two lines are different; hence, the cyclists never have the same velocity.

Part (c): At the intersection of lines A and B, both cyclists are at the same location at the same time. At this point, cyclist A is passing cyclist B, since cyclist A started closer to the origin and has a greater velocity.

REFLECT These three questions show only a small part of what can be learned from the graphs. Graphs offer a parallel representation of physical phenomena, and their interpretation is an important tool in physics and, indeed, science in general.

3: Interpreting a position-versus-time graph

Figure 2.3 shows a position-versus-time graph for the motion of a car. Describe the velocity and acceleration during segments *OA*, *AB*, and *BC*.

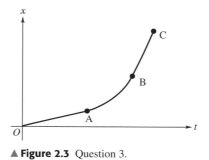

▲ **Figure 2.3** Question 3.

Solution

SET UP AND SOLVE Velocity is the change in position with respect to time, and acceleration is the change in velocity with respect to time. We find the velocity by looking at the slope of the position-versus-time graph and the acceleration by noting how the velocity changes.

In the segment *OA*, the slope is constant, indicating that the velocity is constant. With constant velocity, there is no acceleration.

In the segment *AB*, the slope is increasing smoothly, indicating that the velocity is increasing. There must be acceleration for the velocity to increase.

In the segment *BC*, the slope is again constant, indicating that the velocity is constant. This velocity is greater in magnitude than the velocity in segment *OA*, since the slope is larger. With constant velocity, there is no acceleration.

REFLECT This question illustrates how we can find the velocity and acceleration from the position-versus-time graph.

4: A falling ball

A ball falls from the top of a building. If the ball takes time t_A to fall halfway from the top of the building to the ground, is the time it takes to fall the remaining distance to the ground equal to, greater than, or smaller than t_A?

Solution

SET UP AND SOLVE We can break the problem up into two segments: the first half and the second half. In the first segment, the falling ball starts with an initial velocity of zero. In the second segment, the ball has acquired velocity, so it has an initial velocity. The time to complete the second segment must be shorter than t_A.

REFLECT If you watch a ball fall, you should be able to see that it spends more time in the first half of the motion than in the second half. We can also look at the equation for a falling body:

$$y = y_0 + v_{0y}t + \tfrac{1}{2}a_y t^2.$$

For the first half of the motion the velocity term is zero, and for the second half it is not zero. Given equal time and equal acceleration, a segment with an initial velocity will cover a larger distance, or cover the same distance in a shorter time.

Problems

1: Throwing a ball upwards

A ball thrown vertically upwards from the edge of a 150-m-tall building falls to the ground 9.5 s after leaving the thrower's hand. Assume that the thrower's hand is 2.0 m above the roof of the building. Find the initial velocity of the ball and the time at which the ball reaches its maximum height.

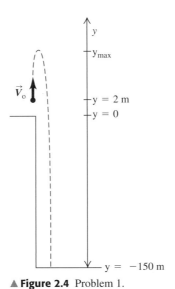

▲ **Figure 2.4** Problem 1.

Solution

SET UP Figure 2.4 shows a sketch of the problem. Once thrown, the ball has an initial velocity and will undergo gravitational acceleration. We will apply the constant-acceleration kinematics equation to solve the problem.

We ignore effects due to the air. A vertical coordinate system is shown in the diagram, with the origin located at the edge of the building and positive values directed upwards.

SOLVE We first determine the initial velocity of the ball. We know the initial and final positions, time, and acceleration of the ball; therefore, we use the equation for position as a function of time:

$$y = y_0 + v_{0y}t + \tfrac{1}{2}a_y t^2.$$

The initial position of the ball (y_0) is $+2.0$ m, the final position (y) is -150 m (the ground is below the edge of the building), the acceleration is $-g$, and the time is 9.5 s. Rearranging terms in order to find the initial velocity v_{0y} gives

$$v_{0y} = \frac{y - y_0 - \frac{1}{2}a_y t^2}{t}.$$

Substituting the values given yields

$$v_{0y} = \frac{(-150\text{ m}) - (2.0\text{ m}) - \frac{1}{2}(-9.8\text{ m/s}^2)(6.5\text{ s})^2}{(6.5\text{ s})} = 8.5\text{ m/s}.$$

The initial velocity of the ball is 8.5 m/s. The value is positive, indicating that the initial velocity is upwards. To find the time required to reach the maximum height, we know that the velocity at the maximum height is momentarily zero, so we can use the equation for velocity as a function of time:

$$v_y = v_{0y} + a_y t.$$

We now solve for the time t when the velocity v_y is zero:

$$t = \frac{v_y - v_{0y}}{a_y} = \frac{(0 - 8.5\text{ m/s})}{(-9.8\text{ m/s}^2)} = 0.87\text{ s}.$$

The ball reaches its maximum height 0.87 s after leaving the thrower's hand.

REFLECT This is a straightforward application of constant-acceleration kinematics. We identified the known and unknown quantities and substituted into appropriate equations to find the unknown quantities.

Practice Problem: Find the maximum height of the ball. *Answer:* $y_{max} = 5.7$ m above the top of the building.

2: Dropping a stone from a moving helicopter

A helicopter is ascending at a constant rate of 18 m/s. Twelve seconds after the helicopter leaves the ground, a stone falls from the helicopter. How long does it take for the stone to reach the ground?

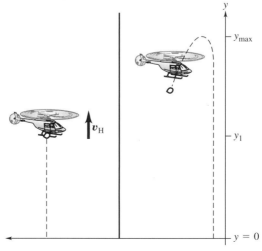

▲ **Figure 2.5** Problem 2.

Solution

SET UP We sketch the problem in Figure 2.5. There are two segments of the stone's motion: moving upward with the helicopter and falling after it breaks free of the helicopter. As it moves upward with the helicopter, the stone has constant velocity. When it falls, the stone has an initial velocity (the same as that of the helicopter) and undergoes acceleration due to gravity. To solve the problem, we will apply the constant-acceleration kinematics equations to the two segments, using the final quantities from the first segment as the initial quantities in the second.

We ignore effects due to the air. A vertical coordinate system is shown in the diagram, with the origin located on the ground and positive values directed upwards.

SOLVE We need to know the position, velocity, and time at the end of the first segment to solve for the second segment. The velocity and time are given in the statement of the problem. The position is found from the equation of position as a function of time with zero acceleration:

$$y = y_0 + v_{0y}t.$$

The initial position is zero (the helicopter and stone start at the ground), and the helicopter is ascending, so v_{y0} is positive 18 m/s and the time is 12 s. Substituting yields

$$y = y_0 + v_{0y}t = 0 + (18 \text{ m/s})(12 \text{ s}) = 216 \text{ m}.$$

For the second segment, the initial position is 216 m, the initial velocity is +18 m/s, the final position is zero, and the acceleration is $-g$. The equation for position as a function of time for constant acceleration can be used to find the time:

$$y = y_0 + v_{0y}t + \tfrac{1}{2}a_y t^2.$$

Substituting gives

$$0 = (216 \text{ m}) + (18 \text{ m/s})t + \tfrac{1}{2}(-9.8 \text{ m/s}^2)t^2.$$

We cannot rearrange this equation to solve directly for t, so we resort to the quadratic equation. (See Problem 1.6.) For this case, $a = -4.9 \text{ m/s}^2$, $b = 18 \text{ m/s}$, and $c = 216 \text{ m}$. The result is given by

$$t = \frac{-b \pm \sqrt{b^2 - 4ac}}{2a}.$$

Substituting and solving yields

$$t = \frac{-(18 \text{ m/s}) \pm \sqrt{(18 \text{ m/s})^2 - 4(-4.9 \text{ m/s}^2)(216 \text{ m})}}{2(-4.9 \text{ m/s}^2)} = -5.1 \text{ s}, \ + 8.7 \text{ s}.$$

The positive solution, 8.7 s, corresponds to the time the stone hits the ground. The stone hits the ground 8.7 s after falling from the helicopter or 20.7 s after the helicopter originally left the ground.

REFLECT We applied the equations of motion with constant acceleration to each of the two segments in this problem, using the results from the first part as input into the second part.

3: Avoiding a ticket

A speed trap is made by placing two pressure-sensitive tracks across a highway, 150 m apart. You notice the speed trap and begin slowing down at the instant you cross the first track. If you are moving at a rate of 42 m/s and the speed limit is 35 m/s, what must your minimum acceleration be in order to have an average speed within the speed limit by the time your car crosses the second track?

Solution

SET UP A sketch of the problem is shown in Figure 2.6. For the average speed over the interval to be under the speed limit, the final speed at the second track must be less than the speed limit. We determine the final speed by finding the average speed in terms of the initial and final speeds. Once we know the final speed, we can find the acceleration from the kinematics equations.

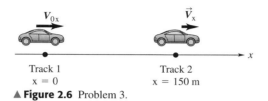

▲ **Figure 2.6** Problem 3.

SOLVE The average speed (for constant acceleration) is

$$v_{\text{av},x} = \frac{v_{0x} + v_x}{2}.$$

Substituting and solving for the final speed gives

$$v_x = 2v_{\text{av},x} - v_{0x} = 2(35 \text{ m/s}) - (42 \text{ m/s}) = 28 \text{ m/s}.$$

The final speed must be 28 m/s for the average speed to be 35 m/s. We can use the equation for velocity as a function of position with constant acceleration:

$$v_x^2 = v_{0x}^2 + 2a_x(x - x_0).$$

In our coordinate system, the difference between the final and initial positions is 150 m. Substituting and solving for the acceleration yields

$$a_x = \frac{v_x^2 - v_{0x}^2}{2(x - x_0)} = \frac{(28 \text{ m/s})^2 - (42 \text{ m/s})^2}{2(150 \text{ m})} = -3.3 \text{ m/s}^2.$$

You will need to accelerate at a rate of -3.3 m/s² to avoid a ticket, with the minus sign indicating that you will need to slow down.

REFLECT The challenge in this problem was to recognize that we needed a final velocity that would result in the correct average velocity. A common mistake is to take the desired *average* velocity as the *final* velocity. Understanding the difference can help you avoid errors (and a ticket!).

4: Changing volume of a sphere

If the surface area of a sphere increases by a factor of 9, by how much does its volume increase?

Solution

SET UP Appendix A of the textbook gives us the relationship among the surface area, volume, and radius of a sphere. We'll find how the radius of the sphere changed as the surface area changed and then use that to result find the corresponding change in volume.

SOLVE The surface area and volume of a sphere are, respectively,

$$A = 4\pi r^2, \qquad V = \tfrac{4}{3}\pi r^3$$

The surface area of a sphere is proportional to the square of the radius. If the surface area increases by a factor of 9, then the radius must have increased by a factor of $\sqrt{9}$, or 3.

The volume of a sphere is proportional to the cube of the radius. If the radius increases by a factor of 3, then the volume will increase by a factor of 3^3, or 27. The volume is 27 times larger than the original volume.

REFLECT The straightforward solution to this problem was to realize that both the surface area and the volume are related to the radius. By understanding how the radius changed, we found how the volume changed. Proportional reasoning is used often in physics to build intuition.

Practice Problem: How would the volume have changed if the surface area decreased by a factor of 4? *Answer:* The volume would have decreased by a factor of 8.

Motion in a Plane

Summary

In this chapter, we expand our kinematics to the motion of objects in a plane, finding that we can simultaneously apply our one-dimensional kinematics equations to two axes independently. Displacement, velocity, and acceleration take on their vector qualities as we expand to a plane, requiring us to work with components of each quantity. Our new skills will allow us to investigate projectile motion and the interesting case of uniform circular motion. By the end of this chapter, you will have a strong foundation for solving kinematics problems and you will be ready to begin our investigation into the causes of motion.

Objectives

- Describe an object's position, velocity, and acceleration in terms of vector quantities.
- Apply equations of motion to objects moving in a plane.
- Describe and analyze the motion of projectiles.
- Analyze an object in uniform circular motion.
- Find the velocities of objects relative to different reference frames moving in two dimensions.

Concepts and Equations

Term	Description
Position Vector	The position vector $\vec{r}$ of an object in a plane is the displacement vector from the origin to that object. It has components x and y.
Average Velocity	An object's average velocity $\vec{v}_{av}$ during a time interval Δt is its displacement $\Delta \vec{r}$ divided by Δt: $$\vec{v}_{av} = \frac{\vec{r}_2 - \vec{r}_1}{t_2 - t_1} = \frac{\Delta \vec{r}}{\Delta t}.$$
Instantaneous Velocity	An object's instantaneous velocity is $$\vec{v} = \lim_{t \to \infty} \frac{\Delta \vec{v}}{\Delta t}.$$
Average Acceleration	The average acceleration $\vec{v}_{av}$ during a time interval Δt is the change in velocity, $\Delta \vec{v}$, divided by Δt: $$\vec{a}_{av} = \frac{\vec{v}_2 - \vec{v}_1}{t_2 - t_1} = \frac{\Delta \vec{v}}{\Delta t}.$$
Instantaneous Acceleration	An object's instantaneous acceleration is $$\vec{a} = \lim_{t \to \infty} \frac{\Delta \vec{v}}{\Delta t}.$$ An object has acceleration if either its speed or direction changes.
Projectile Motion	An object undergoes projectile motion when it is given an initial velocity and then follows a path determined entirely by the effect of a constant gravitational force. The path, or trajectory, is a parabola in the x-y plane. The projectile's vertical motion is independent of its horizontal motion. The horizontal acceleration is zero and the vertical acceleration is $-g$. For a projectile with an initial velocity of magnitude v_0 and direction θ_0 (measured with respect to the ground), its coordinates and velocities as a function of time are $$x = (v_0 \cos\theta_0)t,$$ $$y = (v_0 \sin\theta_0)t - \tfrac{1}{2}gt^2,$$ $$v = v_0 \cos\theta_0,$$ $$v_y = v_0 \sin\theta_0 - gt.$$
Uniform Circular Motion	A particle moving in a circular path of radius R and with constant speed v is said to move in uniform circular motion. The particle has an acceleration of magnitude $$a_{rad} = \frac{v^2}{R},$$ directed toward the center of the circle.
Relative Velocity	When an object P moves relative to a reference frame B, and B moves relative to a second reference frame A, the velocity of P relative to B is denoted by $\vec{v}_{P/B}$, the velocity of P relative to A is denoted by $\vec{v}_{P/A}$, and the velocity of B relative to A is denoted by $\vec{v}_{B/A}$. These velocities are related by the formula $$\vec{v}_{P/A} = \vec{v}_{P/B} + \vec{v}_{B/A}.$$

Conceptual Questions

1: Velocity and acceleration at the top of a projectile's path

A projectile is launched with initial x and y velocities. (a) Describe the velocity and acceleration just before the ball reaches the top of its trajectory. (b) Describe the velocity and acceleration at the instant the ball reaches the top of its trajectory. (c) Describe the velocity and acceleration just after the ball reaches the top of its trajectory.

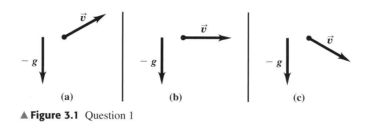

▲ **Figure 3.1** Question 1

Solution

SET UP AND SOLVE Figure 3.1 shows the three time frames we will examine. During the flight, the projectile undergoes acceleration due to gravity. The projectile's initial velocity has both x and y components; the x component remains constant, while the y component is accelerated by gravity. We have to consider both components of velocity separately.

Part (a): The x component of velocity is constant and to the right, and the y component of velocity is upward and very small just before the top of the flight. The acceleration due to gravity is downwards.

Part (b): The x component of velocity is constant and to the right, and the y component of velocity is zero at the top of the flight. The acceleration due to gravity remains constant and downwards.

Part (c): The x component of velocity is constant and to the right, and the y component of velocity is downwards and very small just after the top of the flight. The acceleration due to gravity remains constant and downwards.

REFLECT The acceleration due to gravity causes a change in velocity during the flight. The ball starts with a velocity, which decreases, drops to a minimum value at the top, and then increases downwards. The acceleration due to gravity is constant throughout the motion. The velocity is changing throughout the motion. Question 1 from Chapter 2 is very similar to this question.

2: Launching a marble off the edge of a table

A marble is launched off the edge of a horizontal table and lands on the floor. Draw the trajectory of the ball from the table to the floor. Draw a second line showing the trajectory if the marble is given a smaller initial velocity. Draw a third line showing the trajectory if the marble is given a larger initial velocity.

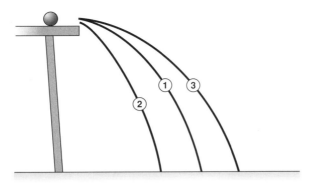

▲ **Figure 3.2** Question 2

Solution

SET UP AND SOLVE The table and marble are sketched in Figure 3.2. The initial trajectory is shown and labeled "1." The marble follows a parabolic path, starting with an initial horizontal velocity. For the smaller initial velocity, the marble also follows a parabolic path, but with a termination point closer to the edge of the table. This is shown on the figure and is labeled "2." For the larger initial velocity, the path follows a parabola, but with a termination point farther from the edge of the table. This is shown on the figure and labeled "3."

REFLECT Each of the paths is similar; their differences owe to the different initial velocities. How does the time spent in the air compare for the three paths? All three take the same amount of time to reach the ground, as they all start with zero initial vertical velocity and fall the same distance. Since they spend the same time in the air, those with larger initial velocities reach greater horizontal distances.

3: Comparing projectiles

Figure 3.3 graphs the paths of two projectiles in the x-y plane. If we ignore air resistance, how do the initial velocities compare (in both magnitude and direction)?

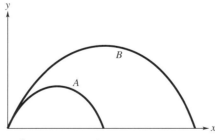

▲ **Figure 3.3** Question 3

Solution

SET UP AND SOLVE At the origin, we see that both paths coincide with each other. This indicates that the directions of the initial velocity of both projectiles are the same.

The graph does not include a time axis, so we need to look to other clues to compare the magnitudes of the projectiles' initial velocities. Trajectory B reaches a greater height, indicating that its initial vertical component of velocity was larger than trajectory A's initial vertical component of velocity. Therefore, trajectory B has the greater initial velocity (magnitude).

REFLECT Without a time axis, we cannot assume that trajectory B has the larger magnitude of initial velocity by examining the x motion.

How does the horizontal component of the initial velocities compare? Both projectiles have the same initial launch angle; therefore, the ratio of their velocity components must be the same. If the vertical component of B's velocity is larger, so must B's horizontal velocity component be larger.

4: Comparing projectiles, part 2

Figure 3.4 graphs the paths of two new projectiles in the *x-y* plane. If we ignore air resistance, how do the initial velocities compare (in both magnitude and direction)? Which projectile lands first?

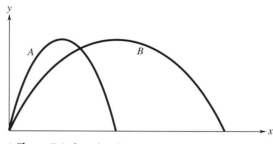

▲ **Figure 3.4** Question 4

Solution

SET UP AND SOLVE We see that the paths do not coincide with each other at the origin. Projectile A has the larger launch angle.

Again, the graph does not include a time axis, so we need to look to other clues to compare the magnitudes of the projectiles' initial velocities. Both trajectories reach the same maximum height, indicating that both have the same vertical velocity components. However, their initial directions were different, requiring their initial horizontal components to be different. Trajectory B reaches a greater horizontal distance, so it must have the larger initial horizontal velocity. Therefore, trajectory B has the greater initial velocity (magnitude).

Since both trajectories reach the same maximum height and have the same initial vertical velocity, they must land at the same time.

REFLECT We could have also considered the time first and the velocity second. In that case, it may have been easier to see that the initial horizontal velocity for projectile B was larger, because it covered more distance in the same time.

Problems

1: Water balloon launch

Your physics professor is walking past the physics building at a constant 3.5 m/s pace. You're on the third floor balcony (25 m above the ground) of the physics building with your new water balloon launcher. The launcher allows you to adjust the speed of the water balloon, but can only launch the balloon horizontally. What launch speed should be set for the water balloon to land on your professor if you launch it just as she passes below? Where will your professor be when the balloon hits her, as measured from a point on the ground directly below you?

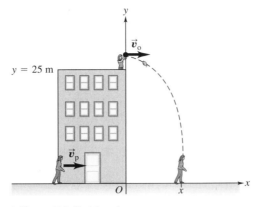

▲ **Figure 3.5** Problem 1

Solution

SET UP Figure 3.5 shows a sketch of the problem. Once launched, the water balloon will undergo gravitational acceleration in the vertical direction and continue with constant velocity in the horizontal direction. We will apply the constant-acceleration kinematics equations separately to the horizontal and vertical components to solve the problem.

We ignore effects due of the air. Your professor is roughly 1.7 m tall, but we'll ignore her height and determine the position where the balloon hits the ground. An x-y coordinate system is shown in the diagram.

SOLVE We first determine the launch speed. Since there is no acceleration in the horizontal direction, the water balloon must be launched at the same speed as your professor is walking, 3.5 m/s.

To find where the balloon hits her, we find the time from the vertical motion and use that to find the horizontal distance the water balloon travels as it falls. The vertical position for constant acceleration is given by the formula

$$y = y_0 + v_{0y}t + \tfrac{1}{2}at^2.$$

We've set the origin at the ground; therefore, the initial position becomes 25 m and the final position becomes 0. The launcher imparts only a horizontal velocity, so the initial vertical velocity is zero. The acceleration is $-g$ since the positive vertical axis is directed upwards. Adding these parameters gives

$$0 = 25 \text{ m} - \tfrac{1}{2}gt^2.$$

Rearranging terms produces

$$t = \sqrt{\frac{2(25 \text{ m})}{(9.8 \text{ m/s}^2)}} = 2.26 \text{ s}.$$

It takes 2.26 s for the balloon to fall to the ground. During this time, it is traveling with constant horizontal velocity. We find the horizontal distance it travels from the formula

$$x = x_0 + v_{0x}t.$$

Your origin is directly below your position on the balcony $(x_0 = 0)$. Substituting the horizontal velocity and time we calculated, we find the horizontal distance to be

$$x = v_{0x}t = (3.5 \text{ m/s})(2.26 \text{ s}) = 7.9 \text{ m}.$$

The water balloon will hit your professor a horizontal distance 7.9 m away from your location.

REFLECT This is a straightforward application of two-dimensional kinematics. We solved for one component of the motion and substituted the result into the other component to arrive at the solution. Note that we solved for the vertical motion and substituted the result into the horizontal motion, the opposite order of the previous problem. Practicing a variety of problems will build proficiency in solving problems involving motion in a plane.

2: Hitting a baseball in Fenway Park

You win a chance to try hitting a baseball over the Green Monster in Fenway Park. The Green Monster is a 37.2-ft- (11.3-m) -high wall in left field of the ball park. The left end is closest to home plate at 310 ft (94.5 m) away. If you give the ball an initial speed of 33 m/s at an initial angle of 47°, by how much does the baseball clear (or miss) the top of the wall?

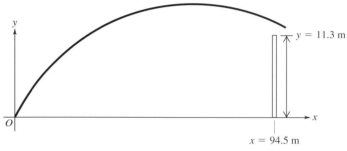

▲ **Figure 3.6** Problem 2 Sketch

Solution

SET UP We sketch the problem in Figure 3.6. The baseball has an initial velocity, undergoes acceleration due to gravity in the vertical direction, and has no acceleration in the horizontal direction. Constant-acceleration kinematics equations will be applied separately to the horizontal and vertical components to find the solution.

We ignore effects due of the air. The ball is hit roughly 1 m or so above the ground, but we'll neglect this small distance and set the origin at ground level. An x-y coordinate system that coincides with this choice is shown in the diagram.

SOLVE We will solve for the time required for the baseball to arrive at the wall, using the horizontal-position equation. Then we will substitute into the vertical-position equation to find the vertical position at the wall. The horizontal position is given by

$$x = x_0 + v_{0x}t.$$

In this case, we start at the origin $(x_0 = 0)$, and the x-component of velocity includes a cosine term:

$$x = v_{0x}t = v_0\cos\theta t.$$

We wish to find the time t when the baseball is located at the wall $(x = 94.5 \text{ m})$:

$$t = \frac{x}{v_0\cos\theta} = \frac{(94.5 \text{ m})}{(33 \text{ m/s})\cos 47°} = 4.20 \text{ s}.$$

After 4.20 s, the baseball's horizontal position is 94.5 m. We now find the vertical position at this time. The vertical position is given by the formula

$$y = y_0 + v_{0y}t + \tfrac{1}{2}at^2.$$

Again, we start at the origin $(y_0 = 0)$, the acceleration is downward (negative) with magnitude g, and the y-component of velocity includes a sine term:

$$y = v_0\sin\theta t + \tfrac{1}{2}(-g)t^2.$$

We can now substitute our values into the equation to find the height:

$$y = (33 \text{ m/s})(\sin 47°)(4.20 \text{ s}) + \tfrac{1}{2}(-9.8 \text{ m/s}^2)(4.20 \text{ s})^2 = 14.9 \text{ m}.$$

At the wall, the height is 14.9 m, or 3.6 m above the 11.3-m-high wall. You clear the Green Monster by 3.6 m!

REFLECT This is another straightforward application of two-dimensional kinematics. We solved for one component of the motion and substituted the result into the other component to arrive at the solution. We will follow this procedure often to solve problems involving motion in a plane.

Practice Problem: Find the x and y components of the baseball's velocity at the wall. *Answer:* $v_x = 22.5 \text{ m/s}, v_y = -17.0 \text{ m/s}.$

3: Acceleration of a propeller tip

The Wright Brothers' plane had a 2.4-m-long propeller that operated at a constant 350 rpm. Find the acceleration of a particle at the tip of the propeller.

Solution

SET UP This is a problem involving uniform circular motion; the acceleration is determined by the centripetal acceleration formula. We will need to find the velocity and radius from the information provided. A diagram of the problem is shown in Figure 3.7.

▲ **Figure 3.7** Problem 3.

SOLVE To find the centripetal acceleration, we need the radius and speed of a particle on the tip of the propeller. We are given the diameter of the propeller; dividing that in half gives the radius. The speed is found by dividing the circumference at the tip of the propeller $(2\pi r)$ by the time it takes to make one revolution (T):

$$v = \frac{2\pi r}{T}.$$

We are given that the propeller makes 350 revolutions per minute, so we find the time it takes to make 1 revolution by dividing 1 minute by 350 revolutions:

$$T = \frac{1 \text{ min}}{350 \text{ rev}} = \frac{60 \text{ s}}{350 \text{ rev}} = 0.171 \text{ s/rev}.$$

The propeller takes 0.171 s to make 1 revolution. We can now find the velocity:

$$v = \frac{2\pi r}{T} = \frac{2\pi(1.2 \text{ m})}{0.171 \text{ s}} = 44.1 \text{ m/s}.$$

The centripetal acceleration is then

$$a_{rad} = \frac{v^2}{r} = \frac{(44.1 \text{ m/s})^2}{(1.2 \text{ m})} = 1,620 \text{ m/s}^2.$$

The centripetal acceleration for a particle on the tip of the propeller is 1,620 m/s². This is equivalent to 165 times the acceleration due to gravity.

REFLECT We have found the magnitude of the acceleration in this problem. The direction of acceleration is towards the center, perpendicular to the velocity.

4: Paddling across a river

You wish to paddle north across a 350-m-wide river. The river has a 1.2 m/s current from east to west and you can paddle at a steady 1.5 m/s pace. In what direction should you paddle and how long will it take you to cross the river?

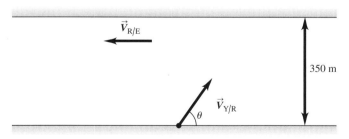

▲ **Figure 3.8** Problem 5

Solution

SET UP Figure 3.8 shows a sketch of the situation. You will need to paddle into the river current to compensate for the river moving your boat downstream as you cross. The direction in which you must paddle is determined by setting the direction of your resulting relative velocity with respect to the earth to be north.

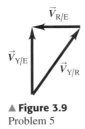

▲ **Figure 3.9**
Problem 5

SOLVE Figure 3.9 combines your velocity with respect to the river $(v_{Y/R})$ with the river current's velocity with respect to the earth $(v_{R/E})$ to form the relative velocity of you moving with respect to the earth $(v_{Y/E})$:

$$\vec{v}_{Y/E} = \vec{v}_{Y/R} + \vec{v}_{R/E}$$

For you to land directly across from your starting point, the direction of $v_{Y/E}$ must be northward. Therefore, the x component of $v_{Y/R}$ must be equal and opposite to $v_{R/E}$. We find the direction you should paddle by equating these two magnitudes:

$$(\vec{v}_{Y/R})_x = v_{R/E},$$
$$(\vec{v}_{Y/R})_x = v_{Y/R}\sin\theta = v_{R/E},$$
$$\theta = \sin^{-1}\left(\frac{v_{R/E}}{v_{Y/R}}\right) = \sin^{-1}\left(\frac{1.2 \text{ m/s}}{1.5 \text{ m/s}}\right) = 53°.$$

You will need to paddle 53° east of north to follow a northward path. The time it will take is found by from the y component of the displacement. You are traveling at constant velocity, so the vertical component of the displacement is

$$y - y_0 = (v_{Y/E})_y t.$$

$(v_{Y/E})_y$ is the magnitude of $v_{Y/E}$, as there is only a y component to this velocity. $(v_{Y/E})_y$ must also be equal to the y component of $v_{Y/R}$. Solving for time yields

$$t = \frac{y - y_0}{(v_{Y/E})_y} = \frac{y - y_0}{v_{Y/E}} = \frac{y - y_0}{(v_{Y/R})_y} = \frac{y - y_0}{v_{Y/E}\cos\theta} = \frac{350 \text{ m}}{(1.5 \text{ m/s})\cos(53°)} = 390 \text{ s}.$$

It will take you 390 s to paddle across the river.

REFLECT When you paddle across a river perpendicular to its flow, your relative velocity with respect to the earth is always less than your velocity with respect to the river. It also takes longer to cross a river with a current compared with a calm river. The next practice problem lets you compare the time required to cross a calm river with the time you just found in treating a river with a current.

Practice Problem: How long would it take to paddle across the same river if no current were present? *Answer:* 230 s.

NEWTON'S LAWS OF MOTION

4

Summary

We will define dynamics, the study of the relationship of motion to forces, in this chapter. Newton's laws of motion will lay the foundation for our studies and link forces to acceleration, which we investigated in earlier chapters. We will define force, mass, and weight and apply them to problems. We will use our knowledge of vectors to better understand forces and construct free-body diagrams. By the end of this chapter, we will have built a problem-solving framework that we will apply in the next chapter.

Objectives

- Identify forces acting on a body.
- Learn and understand Newton's three laws of motion.
- Find the resultant force by summing multiple forces.
- Recognize an inertial frame of reference in which Newton's laws are valid.
- Use a free-body diagram to represent forces acting on an object.
- Use a free-body diagram as a guide in writing force equations for Newton's laws.

Concepts and Equations

Terms	Descriptions
Force	A force is a quantitative measure of the interaction between two objects. The SI unit of force is the newton (N). One newton equals 1 kilogram-meter per second squared.
Contact Force	A contact force is a force describing the interaction between two objects touching at a surface. A contact force has two components: a component perpendicular to the surface (the normal force) and a component parallel to the surface (the frictional force).
Normal Force	The normal force, denoted by $\vec{n}$, is the component of contact force between two objects that is perpendicular to the surface.
Frictional Force	The frictional force, denoted by $\vec{f}$, is the component of contact force between two objects parallel to the surface. Frictional forces often act to resist the sliding of an object.
Tension Force	A tension force, denoted by $\vec{T}$, is conveyed by the pull of a rope or cord.
Weight	An object's weight, denoted by $\vec{w}$, is the gravitational force exerted on the object by the earth or another astronomical body.
Resultant	The vector sum of the forces acting on a body is the resultant, denoted $\vec{R}$. The effect of many forces acting on a body can be replaced by the resultant according to a principle called **superposition of forces.**
Inertial Reference Frame	An inertial reference frame is a reference frame in which Newton's laws are valid. A common example of a *non*inertial reference frame is that of an accelerating airplane.
Newton's First Law	Newton's first law states that an object will remain at rest or move with constant velocity unless it is acted upon by a net force. The object is said to be in equilibrium if there are no net forces acting on it. The law is valid only in inertial reference frames.
Newton's Second Law	Newton's second law of motion states that an object that is not in equilibrium is being acted upon by a net force and accelerates. The acceleration is given in vector form by $$\sum \vec{F} = m\vec{a}$$ where m is the object's mass. It can also be written in component form as $$\sum F_x = ma_x \quad \text{and} \quad \sum F_y = ma_y$$
Newton's Third Law	Newton's third law states that each of two interacting objects A and B exerts a force on the other that is of equal magnitude and in opposite direction, or $$\vec{F}_{A \text{ on } B} = -\vec{F}_{B \text{ on } A}.$$
Free-Body Diagram	A free-body diagram is a diagram showing all forces acting **on** an object. The object is represented by a point; the forces are indicated by vectors. A free-body diagram is useful in solving all problems involving forces.

Conceptual Questions

1: Winning a tug-of-war match

In a tug-of-war match shown in Figure 4.1, how does the force applied by the losing team to the rope compare with the force applied by the winning team to the rope? Is the magnitude of the force applied by the losing team less than, greater than, or equal to the magnitude of the force applied by the winning team? How does the winning team win?

▲ **Figure 4.1** Question 1.

Solution

SET UP AND SOLVE The free-body diagram in Figure 4.2 will guide our analysis. Each team experiences four forces: the tension due to the tug-of-war rope (T), a frictional force (f), the combined weight of the team members (W), and the normal force from the ground (n). The subscripts indicate the winning (w) and losing (l) teams. Examining the diagrams, we see that the tension forces must be an action-reaction pair—hence the explicit notation. Therefore, by Newton's third law, the tensions must be equal. The force applied by the losing team on the rope must be of the same magnitude as, but opposite in direction to, the force applied by the winning team on the rope.

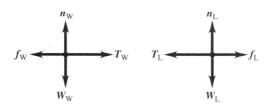

▲ **Figure 4.2** Question 1 free-body diagrams.

The vertical forces will not influence the horizontal interaction, so we look at the remaining force to determine how the winning team wins. The frictional forces must not be equal; the winning team exert a larger frictional force than the losing team in order to accelerate the losing team across the centerline.

REFLECT We see that free-body diagrams and Newton's third law were crucial in our solution. The free-body diagram helped reduce the complexity of the problem and helped show that the frictional force was responsible for the win. After establishing that the tensions were an action-reaction pair, Newton's third law showed us that the tensions were equal and opposite.

Note that we assumed that the frictional force was between the team and the ground and that each team was able to grip the rope without sliding. There is also a frictional force between the teams' hands and the rope. Differences between the teams' hand–rope frictional forces could also have led to the win.

2: Flying groceries

What force causes a bag of groceries to fly forward when you come to an abrupt stop in a car?

Solution

SET UP AND SOLVE Suppose that before you come to an abrupt stop, you are moving at a constant velocity. Then no net force must act on you, the car, or the bag of groceries, according to Newton's first law. As you cause an abrupt stop by hitting the brakes, you increase the frictional force between the car and the road, creating a net force on the car. When you brake, the force on the bag of groceries doesn't change, so the bag of groceries continues at its initial velocity. (We're assuming that the frictional force between the bag of groceries and the car seat is small.) Therefore, *no force* causes the bag of groceries to fly forward when you come to an abrupt stop in a car.

REFLECT The solution may seem a bit illogical, for consider how the situation would appear to someone outside of the car: The bag of groceries continues moving at a constant velocity after the brakes are applied. This scenario should be more plausible and is a clearer way to imagine the situation.

   This is one example of a noninertial frame of reference. The slowing car has negative acceleration and hence is an accelerated frame of reference. Newton's laws don't apply to noninertial frames of reference, so we cannot apply our new force techniques to this problem.

   From inside the car, you may try to explain the situation by invoking a "force of inertia." This would be a fictitious force, however, and should be avoided. All of the forces we've encountered (and all of those we will encounter later) arise from known interactions.

3: Does an apple accelerate when placed on a table?

An apple is placed on a table. Can we describe the apple as having an acceleration of 9.8 m/s^2 towards earth and a second acceleration of 9.8 m/s^2 upwards due to the table, thus resulting in a net acceleration of zero?

Solution

SET UP AND SOLVE We have seen how forces can cause accelerations, have heard $F = ma$ often, and know that an object's weight is *mg*, so it may appear logical to replace forces with mass times acceleration in equations. However, Newton's laws apply to combining *forces,* not accelerations. Newton's second law states that a net force on an object will lead to an acceleration equal to the net force divided by the object's mass.

REFLECT This question shows a common misconception about accelerations and forces. At times, replacing forces with mass times acceleration may lead to the same results as following the correct procedures, but doing so often leads to confusion. An object that is stationary is not accelerating, because there is no *net* force.

4: Forces and moving objects

Does a force cause an object to move? Does a moving object have a force?

Solution

SET UP AND SOLVE A force does not necessarily cause an object to move. Your textbook is acted upon by gravity when it is placed on a desk, but it does not move. A net force can cause an object to acquire velocity through acceleration.

An object moving at constant velocity has no net force acting on it; therefore, motion does not require a force. An *accelerating* object would certainly indicate at least one force acting upon it.

REFLECT Acceleration and motion are *not* equivalent. Acceleration is motion during which the velocity changes over time. An object can also have a *constant* velocity, which would be motion without acceleration. You must distinguish carefully between motion and acceleration to learn physics well.

5: Definition of equilibrium

Can a moving object be in equilibrium?

Solution

SET UP AND SOLVE Equilibrium occurs when the net force on an object is zero. Newton's first law states that objects with no net force acting on them remain at rest or continue moving with constant velocity. An object moving at constant velocity is in equilibrium.

REFLECT Equilibrium has a precise definition in physics, even though the word may have connotations of a stationary object. Physics relies upon precise definitions to build the representations of physical processes. You must apply physics definitions carefully to build your physics understanding.

Problems

1: Combining several forces to find the resultant

A mover uses a cable to drag a crate across the bed of the mover's truck as shown in Figure 4.3. The mover provides a 300 N force and pulls the cable at an angle of 30.0°. The crate weighs 500 N, and the truck bed provides a 350 N normal force on the crate and opposes the mover's pull with a 150 N frictional force. Find the resulting force acting on the crate. Will the crate accelerate?

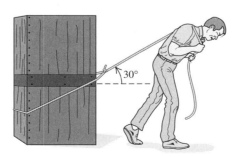

▲ **Figure 4.3** Problem 1.

Solution

SET UP We find the resultant force by adding the forces acting on the crate. Four forces act on the crate: the tension force due to the mover's pull (T), the crate's weight (W), and the normal force (n) and friction force (f) due to the truck bed. We represent the four forces as vectors in the free-body diagram of the crate in Figure 4.4.

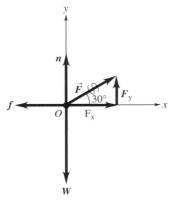

▲ **Figure 4.4** Problem 1 free-body diagram.

We have added an x-y coordinate system to the free-body diagram because the forces act in two dimensions. We've also resolved the tension force into its x and y components.

SOLVE We add the four forces together by adding their components, writing separate equations for the x and y components. There are two x components, due to the horizontal component of the tension force and the friction force:

$$\sum F_x = T\cos 30^\circ + (-f).$$

The x component of the tension force is directed to the right and is thus assigned a positive value, while the friction force is directed to the left and is assigned a negative value. We proceed to the y components. There are three y components, due to the normal force, the weight of the crate, and the vertical component of the tension force:

$$\sum F_y = n + (-W) + T\sin 30^\circ.$$

The y component of the tension force and the normal force are directed upwards and are assigned positive values, while the weight is directed downwards and is assigned a negative value. We now substitute the values for the variables to find the net force along both axes:

$$\sum F_x = T\cos 30^\circ + (-f) = (300\,\text{N})\cos 30^\circ + (-150\,\text{N}) = +110\,\text{N},$$
$$\sum F_y = n + (-W) + T\sin 30^\circ = (350\,\text{N}) + (-500\,\text{N}) + (300\,\text{N})\sin 30^\circ = 0\,\text{N}.$$

The resultant force on the crate has an x component of $+110$ N and no y component (i.e., the resulting force is horizontal and points to the right). There is a resulting acceleration of the crate to the right, because a net force acts on it.

REFLECT This is a typical force problem in which we have used our vector addition skills to find the resultant force. We see that there is no resulting force in the vertical direction; therefore, the crate remains on the truck bed.

Practice Problem: At what rate does the crate accelerate? *Answer:* 2.2 m/s².

2: Using Newton's second law to find the mass of a cruise ship

A tugboat pulls a cruise ship out of port, as shown in Figure 4.5. You estimate the ship's acceleration by noting the tugboat takes 60 s to move the ship 100 m, starting from rest. If the tugboat exerts 3×10^6 N of thrust, what is the mass of the cruise ship? Ignore drag due to the water, and assume that the tugboat accelerates uniformly.

▲ **Figure 4.5** Problem 2.

Solution

SET UP Newton's second law can be used to find the mass of the cruise ship if we know the net force on it and its acceleration. The problem gives the net force provided by the tugboat, and the acceleration can be determined from the kinematics information. We ignore drag and friction with the water, so the only horizontal force acting on the cruise ship is due to the tugboat.

SOLVE Newton's second law relates the net force to the mass and resulting acceleration:

$$\sum F_x = ma.$$

The net force acting on the cruise ship is 3×10^6 N. The acceleration is found from the equation for position as a function of time with constant acceleration:

$$x = x_0 + v_0 t + \tfrac{1}{2}at^2.$$

Here, the initial velocity is zero and we take the initial position to be zero. Substituting these values into the equation then gives

$$x = \tfrac{1}{2}at^2.$$

Rearranging terms to solve for acceleration yields

$$a = \frac{2x}{t^2}.$$

Replacing the distance and time with the given values produces

$$a = \frac{2x}{t^2} = \frac{2(100 \text{ m})}{(60 \text{ s})^2} = 0.056 \text{ m/s}^2.$$

We now use Newton's second law to find the mass. Rearranging terms gives

$$m = \frac{F}{a} = \frac{(3 \times 10^6 \text{ N})}{(0.056 \text{ m/s}^2)} = 54{,}000{,}000 \text{ kg} = 54 \text{ kilotonnes}$$

Our estimate shows that the cruise ship has a mass of 54 kilotonnes (1 kilotonne $= 10^6$ kg). More correctly, the cruise ship has a mass of 50 kilotonnes, as the values given in the statement of the problem have only one significant figure.

REFLECT This problem shows how we can combine Newton's law with observations to make interesting conclusions about the mass of an object.

3: Drawing free-body diagrams

Draw a free-body diagram for each of the following cases:

(a) A box slides down a smooth ramp. (See Figure 4.6).

▲ **Figure 4.6** Problem 3a

(b) A box slides down a rough ramp. (See Figure 4.7).

▲ **Figure 4.7** Problem 3b

(c) A block is placed on top of a crate and the crate is placed on a horizontal surface. (See Figure 4.8). Draw a free-body diagram of the crate.

▲ **Figure 4.8** Problem 3c

(d) A block is placed on top of a crate and the crate is pulled horizontally across a rough surface. (See Figure 4.9). The surface between the crate and block is rough, and the block is held at rest by a string. Draw a free-body diagram of the crate.

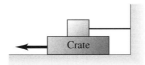

▲ **Figure 4.9** Problem 3d

Solution

SET UP A free-body diagram shows all of the vector forces acting on an object. The first step is to identify the object and then find the forces acting on it. We'll look at the contact tension, normal, and frictional forces and the noncontact gravitational force.

SOLVE In part (a), there is no friction, since the ramp is smooth. The only contact force acting on the box is the normal force due to the ramp. Gravity also acts on the box. The free-body diagram includes two forces acting on the box: the normal (n) force perpendicular to the ramp and the weight (W) downward. The free-body diagram of the box is shown in Figure 4.10.

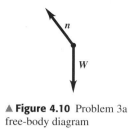

▲ **Figure 4.10** Problem 3a free-body diagram

In part (b), there is friction, as the ramp is rough. The contact forces acting on the box are the normal and frictional forces due to the ramp. Gravity also acts on the box. The free-body diagram includes three forces acting on the box: the normal force (n) perpendicular to the ramp, the frictional force (f) upward along the ramp (opposing the motion), and the weight (W) downward. The free-body diagram of the box is shown in Figure 4.11.

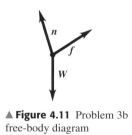

▲ **Figure 4.11** Problem 3b free-body diagram

In part (c), two contact forces act on the crate: the normal force due to the surface and the normal force due to the block. There are no frictional forces, as neither the crate nor the block is moving. Gravity acts on the box. The free-body diagram includes three forces acting on the crate: the normal force due to the surface (n_{surface}), directed upward; the normal force due to the block (n_{block}), directed downward; and the weight (W), directed downward. The free body diagram of the crate is shown in Figure 4.12.

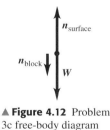

▲ **Figure 4.12** Problem 3c free-body diagram

In part (d), five contact forces act on the crate: the normal forces due to the surface and the block, the frictional forces due to the surface and the block, and the tension force provided by the pull. The sixth force acting on the crate is gravity. The free-body diagram includes six forces acting on the crate: the normal force due to the surface (n_{surface}), directed upward; the normal force due to the block (n_{block}), directed downward; the frictional forces due to the surface (f_{surface}) and the block (f_{block}), both directed to the right; the tension force (T), directed to the left; and the weight (W), directed downward. The free-body diagram of the crate is shown in Figure 4.13.

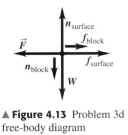

▲ **Figure 4.13** Problem 3d
free-body diagram

4: Tension in a string connecting two blocks

Two blocks are connected by a massless string. A cable is attached to the upper block and is pulled upward with a 250 N force. Find the tension in the string. The upper block has a mass of 7.5 kg and the lower block has a mass of 12 kg.

▲ **Figure 4.14**
Problem 4

Solution

SET UP We will use one of Newton's laws to solve this problem. From the information given, we cannot determine whether the system is in equilibrium or accelerating; therefore, we do not know whether to apply Newton's first law for a system in equilibrium or Newton's second law for an accelerating system. Our first step will be to determine whether the system is in equilibrium or is accelerating. Then we apply Newton's law to find the tension in the string.

We'll use free-body diagrams to solve the problem. We can determine whether the blocks are accelerating by considering the two blocks as one system. The left side of Figure 4.15 shows a free-body diagram of the system with the two blocks combined. We can find the tension in the string by considering the two blocks separately. The right side of Figure 4.15 shows the free-body diagrams for the two blocks separately. The top block is designated "A," the bottom block "B," to reduce confusion.

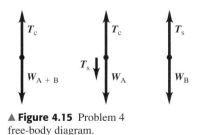

▲ **Figure 4.15** Problem 4
free-body diagram.

The forces are identified by their magnitudes in the free-body diagrams. The combined diagram includes the tension of the cable (T_{cable}) and the weight of the two blocks (W_{A+B}). The other diagrams also include the tension in the string (T_{string}) and the weights of the blocks (W_A and W_B). The upward-pointing vectors will be taken to be positive, the downward-pointing vectors negative.

SOLVE To determine whether the blocks are accelerating, we examine the net force acting on them. From the left-hand free-body diagram, we see that there are two forces acting on the combination of blocks: the tension of the cable and the combined weights of the blocks. Thus,

$$\sum F_y = T_{cable} + (-W_{A+B}).$$

The weight is the combined masses of the blocks times the gravitational constant. The net force is found by replacing the weight and tension by the given values:

$$\sum F_y = T_{cable} + (-g(m_A + m_B)) = 250\,\text{N} + (-(9.8\,\text{m/s}^2)(7.5\,\text{kg} + 12\,\text{kg})) = 58.9\,\text{N}.$$

The net force is nonzero; therefore, the blocks are accelerating. We find the acceleration from Newton's second law applied to the combined blocks:

$$\sum F_y = (m_A + m_B)a.$$

Solving for acceleration yields

$$a = \frac{\sum F_y}{(m_A + m_B)} = \frac{58.9\,\text{N}}{(7.5\,\text{kg} + 12\,\text{kg})} = 3.02\,\text{m/s}^2.$$

We can now apply Newton's second law to the lower block to find the tension in the string. Two forces are acting on the lower block: the tension due to the string (upward) and gravity (downward). Hence,

$$\sum F_y = T_{string} + (-m_B g) = m_B a.$$

Solving for the tension in the string and substituting the value for the acceleration gives

$$T_{string} = m_B g + m_B a = m_B(g + a) = (12\,\text{kg})(9.8\,\text{m/s}^2 + 3.02\,\text{m/s}^2) = 150\,\text{N}.$$

The tension in the string is 150 N.

REFLECT We see that the tension force due to the string is less than the tension force due to the cable. This is expected, as the string provides force to accelerate the lower block while the cable provides force to accelerate both blocks. It is important not to assume that tensions are equal in problems; you must consider each cable independently.

5

APPLICATIONS OF NEWTON'S LAWS

Summary

We will apply Newton's laws of motion to objects that are in *equilibrium* (at rest or in uniform motion) and to objects that are *not in equilibrium* (in accelerated motion) in this chapter. We'll develop a consistent problem-solving strategy that utilizes a free-body diagram to identify the relevant forces acting on an object. Identifying these forces will allow us to quantify them and solve for the object's motion or unknown force. We'll also expand our catalog of forces by quantifying contact forces, frictional forces, and elastic forces. By the end of the chapter, you will have built a foundation for solving equilibrium and nonequilibrium problems involving any combination of forces, including those we'll discover in later chapters.

Objectives

- Represent forces acting on an object with a free-body diagram.
- Use the free-body diagram as a guide in writing force equations for Newton's laws.
- Sum the individual forces acting on an object to zero for objects in equilibrium.
- Write the equations of motion for objects not in equilibrium.
- Apply Newton's laws to two-dimensional problems.
- Understand contact forces, frictional forces, and elastic forces and apply them to a variety of situations.
- Become proficient at using Newton's laws to solve problems.

Concepts and Equations

Term	Description
Free-Body Diagram	A free-body diagram is a diagram showing all forces acting **on** an object. The object is represented by a point; forces are indicated by vectors. A free-body diagram is useful in solving all problems involving forces.
Object in Equilibrium	An object in equilibrium (either at rest or moving with constant velocity) has no net force acting on it; the vector sum of the forces acting on the object must be zero according to Newton's first law of motion: $\Sigma \vec{F} = 0$. When solving equilibrium problems, one starts with the free-body diagrams and finds the net forces along two perpendicular components: $$\sum F_x = 0, \qquad \sum F_y = 0.$$
Applying Newton's Second Law	An object that is not in equilibrium is acted upon by a net force and accelerates. The acceleration is given by Newton's second law of motion: $$\sum \vec{F} = m\vec{a}.$$ When solving nonequilibrium problems, one starts with the free-body diagrams, finds the net forces along two components, and writes down the equations of motion: $$\sum F_x = ma_x, \qquad \sum F_y = ma_y.$$ In circumstances involving more than one object, it may be necessary to apply Newton's laws to each object individually and solve the equations simultaneously.
Frictional Force	A frictional force is that component of the contact force between two objects which is parallel to the surfaces in contact. Frictional forces, denoted by $\vec{f}$, are generally proportional to the normal force and include kinetic-frictional forces (in cases where there is motion between the two objects), static-frictional forces (in cases where there is no motion between the objects), viscosity and drag forces (for motion involving liquids and gases), and rolling-frictional forces (for rolling objects).
Kinetic-Frictional Force	A kinetic-frictional force is a frictional force between two moving objects that is generally proportional to the normal force between the objects. The proportionality constant is the coefficient of kinetic friction (μ_k), which depends on the objects' surface characteristics and has no units. The direction of the kinetic-frictional force is always opposite to the direction of motion. Mathematically, $$f_k = \mu_k n.$$
Static-Frictional Force	A static-frictional force is a frictional force between two objects that are not moving relative to each other. The maximum frictional force is generally proportional to the normal force between the objects, where the proportionality constant is the coefficient of static friction (μ_s). Often, μ_s is greater than μ_k for a given pair of surfaces. The static-frictional force can vary from zero to the maximum value; its magnitude depends on the component of the applied forces parallel to the surface. The direction of the static-frictional force is opposite that of the parallel component of the net applied force. Mathematically, $$f_s \leq \mu_s n.$$

Elastic Force	An elastic force is a force that restores a solid object to its original equilibrium position after deformation. For a spring, the deformation is approximately proportional to the applied force, as given by Hooke's law; that is, $$F_{spr} = -kx,$$ where k is the force constant, x is the displacement of the spring from its equilibrium position, and the negative sign indicates that the force is in the direction opposite to the displacement (i.e., the force is in the direction that would restore the spring to its unstretched position).

Conceptual Questions

1: Finding errors in a free-body diagram

Two weights are suspended from the ceiling and each other by ropes as shown in Figure 5.1a. A free-body diagram is shown in Figure 5.1b for the upper block (A). Find the error in the free-body diagram and draw the correct diagram.

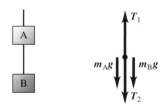

▲ **Figure 5.1** Question 1.

Solution

SET UP AND SOLVE Three forces act on block A: two tension forces due to the ropes and the gravitational force on block A. The gravitational force on block B has been incorrectly included in the diagram. Block B is not in direct contact with block A; only the rope is in contact with block A. The corrected free-body diagram is shown in Figure 5.2.

▲ **Figure 5.2** Question 1 corrected free-body diagram.

REFLECT When drawing free-body diagram, one must include only those forces acting **on** the object. Identifying the forces acting on an object is necessary to apply Newton's laws successfully.

2: Normal force investigation

For which of the following figures is the normal force not equal to the object's weight?

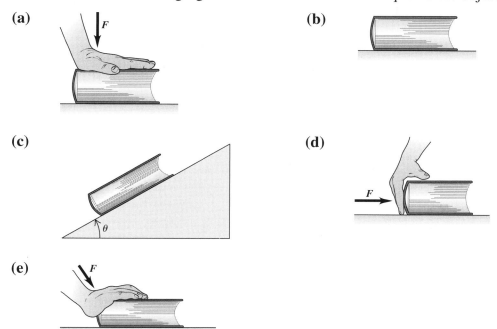

(a)

(b)

(c)

(d)

(e)

Solution

SET UP AND SOLVE The normal force is not equal to the object's weight in figures (a), (c), and (e). In figure (a), the normal force is equal to the book's weight plus the force pushing down on the book. In figure (c), the book's weight is directed downwards and the normal force is directed up and to the left, perpendicular to the ramp's surface. Here, the normal force is equal to the weight multiplied by the cosine of the ramp angle. In figure (d), the normal force is equal to the book's weight and the applied force is along the surface; thus, it does not affect the vertical forces. In figure (e), there is a component of the applied force parallel to the normal force, thus increasing it by the applied force multiplied by the sine of the angle.

REFLECT Often, the normal force is not equal to an object's weight. A common mistake initially encountered in force problems is assuming that the normal force is always equal to some object's weight. You must analyze all problems carefully to determine the proper normal force.

3: Acceleration and tension in blocks connected by a rope

For the following two questions, consider the situation shown in Figure 5.3. Cart A is placed on a table and connected to block B by a rope that passes over a frictionless pulley.

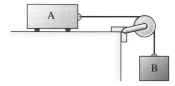

▲ **Figure 5.3** Question 3.

(a) How does the acceleration of cart A compare with that of block B?

Solution

SET UP AND SOLVE Both objects must accelerate at the same rate, since they are connected by the rope (as long as the rope doesn't stretch). To get a better intuition for this situation, note that cart A will move 10 cm when block B moves 10 cm. If block B moves the 10 cm in 1 second, then cart A moves 10 cm in 1 second; their velocities are the same. If block B's velocity changes by 2 m/s in 1 second, then block A's velocity must change by 2 m/s in 1 second; their accelerations are the same. We say that the rope constrains both objects to accelerate at the same rate.

(b) How does the tension force acting on cart A compare with the weight of block B as the system accelerates?

Solution

SET UP AND SOLVE The tension force in the string is constant along the length of the string, so the tension force is the same on block B as it is on cart A. Therefore, we compare the tension at block B with block B's weight. Newton's second law tells us that the net force on an object is equal to its mass times its acceleration. Two forces act on block B: B's weight and the tension force. The weight minus the tension force must be equivalent to the acceleration multiplied by block B's mass; therefore, the tension must be less than the weight. The tension force on cart A is less than the weight of block B.

REFLECT Solving this problem gives us two important results that we'll apply repeatedly to later problems. First, objects connected by a rope are constrained to have the same magnitude of acceleration. Second, the tension force in a rope connected to an object is not always equal to the object's weight if the object is accelerating.

4: What can a hanging ball indicate?

A ball hangs on a string attached to the top of a box, as shown in Figure 5.4. You observe the ball and find that it remains in the position shown in the figure for a long time. By looking only inside the box, what can be learned about the motion of the box?

▲ **Figure 5.4**
Question 4.

Solution

SET UP AND SOLVE We can see that the ball has swung to the left, so we may suspect that the box is moving. Let's look at the forces acting on the box to investigate its motion. Two forces act on the ball: gravity and the tension force due to the string. Figure 5.5 shows the free-body diagram.

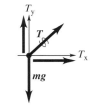

▲ **Figure 5.5** Question 4
free-body diagram

We see that the tension force has components in both the vertical and horizontal directions. The vertical component of the tension force appears equivalent to the gravity force, so we conclude that there is no vertical acceleration. There is only one horizontal force—the horizontal component of the tension force—so there must be horizontal acceleration to the right.

We conclude that the box is accelerating to the right in the horizontal direction and is not accelerating in the vertical direction. We cannot determine the velocity of the box; both the vertical and horizontal velocity components could be zero or nonzero. For example, the box could be moving on a flat surface to the left with decreasing velocity, or it could be accelerating to the right from rest.

REFLECT This problem shows that no net force results in zero acceleration for an object, and a net force results in acceleration for an object. We cannot determine the velocity of an object by knowing only its acceleration; we need additional information.

How would the ball appear if the box was moving with constant velocity? *Answer:* The ball would hang vertically, since no net force would act on it.

5: Motion of a box on a rough surface

A constant horizontal force is applied to a box on a rough floor. With a 15 N force applied, the box begins to slide. What is the motion of the box after it begins to slide, assuming that the applied force remains constant?

Solution

SET UP AND SOLVE Before the box slides, there is a static-frictional force. Once it begins to slide, the static-frictional force becomes a kinetic-frictional force. The kinetic-frictional force is smaller in magnitude than the static-frictional force; therefore, the applied force is larger than the frictional force, and the box accelerates.

REFLECT This problem helps illustrate the fact that the kinetic-frictional force is generally less than the static-frictional force. The reason is that the coefficient of kinetic friction is often less than the coefficient of static friction.

6: Frictional forces

A box is placed on a rough floor. When you push horizontally against the box with a 20 N force, the box just begins to slide. What is the magnitude of the frictional force when you push against the box with a 10 N force? With a 15 N force?

Solution

SET UP AND SOLVE Since the box just begins to slide when the 20 N force is applied, the maximum static-frictional force is 20 N. When you push against the box with a 10 N force, you are pushing with less than the maximum static-frictional force. By Newton's third law, the box must push back with the same force; therefore, the static-frictional force must have a magnitude of 10 N. For the same reason, when you push with a 15 N force, the static-frictional force has a magnitude of 15 N.

REFLECT The static-frictional force varies from zero to its maximum value. The static-frictional force equals the net force acting against the frictional force. Be careful not to assume that the static-frictional force is always at its maximum.

7: A vertical frictional force

A block is placed against the vertical front of an accelerating cart as shown in Figure 5.6. What condition must hold in order to keep the block from falling?

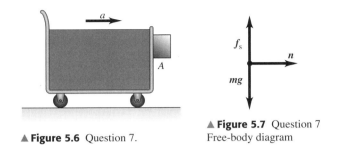

▲ **Figure 5.6** Question 7.

▲ **Figure 5.7** Question 7
Free-body diagram

Solution

SET UP AND SOLVE The free-body diagram shown in Figure 5.7 indicates that three forces act on the box: gravity (mg downwards), the normal force due to the cart (n, to the right), and the frictional force at the block–cart surface (f_s). To keep the block from falling, the frictional force must be equal and opposite to the gravitational force. The condition for equilibrium in the vertical direction gives

$$\sum F_y = f_s - mg = 0, \quad f_s = mg.$$

The frictional force must be the static-frictional force in order to prevent the block from moving. The static-frictional force is given by

$$f_s \leq \mu_s n.$$

The gravitational force can then be related to the normal force and the coefficient of static friction:

$$mg \leq \mu_s n.$$

The block accelerates to the right with acceleration a, so we can use Newton's second law to find an expression for the normal force:

$$\sum F_x = n = ma.$$

Replacing the normal force with mass and acceleration gives

$$mg \leq \mu_s ma, \quad g \leq \mu_s a.$$

Thus,

$$\mu_s \geq \frac{g}{a}.$$

This tells us that the coefficient of static friction must be equal to or greater than the gravitational constant divided by the cart's acceleration.

REFLECT Problems that initially appear complicated often have relatively straightforward solutions. Following a consistent problem-solving procedure helps identify the key points that you will need to solve the problem.

Problems

1: Equilibrium in two dimensions

A 322 kg block hangs from two cables as shown in Figure 5.8. Find the tension in cables A and B.

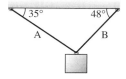

▲ **Figure 5.8** Problem 1.

Solution

SETUP The two cables hold the block in place, so we'll look at the forces acting on the block. Three forces act on the block: the two tension forces $(T_A$ for cable A and T_B for cable B) and gravity (mg). We represent the three forces as vectors in the free-body diagram of the block shown in Figure 5.9.

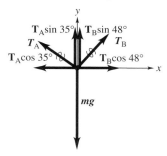

▲ **Figure 5.9** Problem 1 free-body diagram.

We have added an x-y coordinate system to the figure because the forces act in two dimensions. We've also resolved the two tension forces into their x and y components.

SOLVE We apply the equilibrium conditions to the block, writing separate equations for the x and y components:

$$\sum F_x = 0, \quad T_B \cos 48° + (-T_A \cos 35°) = 0,$$
$$\sum F_y = 0, \quad T_B \sin 48° + T_A \sin 35° + (-mg) = 0.$$

Note that components directed to the left and downward are negative, consistent with our coordinate system. We can rewrite the first equation as

$$T_B = T_A \frac{\cos 35°}{\cos 48°}.$$

Substituting for T_B in the second equation gives

$$T_A \frac{\cos 35°}{\cos 48°} \sin 48° + T_A \sin 35° - mg = 0.$$

We can rearrange terms to find T_A:

$$T_A \left(\frac{\cos 35°}{\cos 48°} \sin 48° + \sin 35° \right) = mg,$$

$$T_A = mg \frac{1}{\left(\dfrac{\cos 35°}{\cos 48°} \sin 48° + \sin 35° \right)} = (322 \text{ kg})(9.80 \text{ m/s}^2) \frac{1}{\left(\dfrac{\cos 35°}{\cos 48°} \sin 48° + \sin 35° \right)} = 2{,}130 \text{ N}.$$

Substituting the value for T_A into the first equation gives

$$T_B = T_A \frac{\cos 35°}{\cos 48°} = 2{,}130 \text{ N} \frac{\cos 35°}{\cos 48°} = 2{,}600 \text{ N}.$$

The tension in cable A is 2,130 N and the tension in cable B is 2,600 N.

REFLECT The sum of the magnitudes of the two tension forces (4,730 N) is larger than the weight of the block (3,160 N). This is consistent, as the tension forces are in two dimensions and their magnitudes are greater than their components. In addition, we check whether the two horizontal components of the tension forces are equal in magnitude. Substituting into the first equation gives a magnitude of 1,740 N for each component, each having opposite sign.

2: Accelerated motion of a block on an inclined plane

A block with mass 3.00 kg is placed on a frictionless plane inclined at 35.0° above the horizontal and is connected to a second hanging block with mass 7.50 kg by a cord passing over a small, frictionless pulley. (See Figure 5.10.) Find the acceleration (magnitude and direction) of the 3.00 kg block.

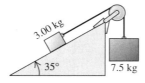

▲ **Figure 5.10** Problem 2.

Solution

SETUP To find the acceleration, we'll apply Newton's second law to the left-hand block. We will also need to apply Newton's second law to the right-hand block and solve the two equations simultaneously to determine the acceleration. The free-body diagrams of both blocks are shown in Figure 5.11.

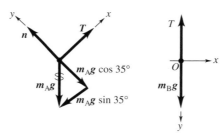

▲ **Figure 5.11** Problem 2 free-body diagrams.

The forces are identified by their magnitudes. Acting on the left-hand block (block A) are gravity $(m_A g)$, the normal force (n), and the tension force (T). The right-hand block (block B) is acted on only by the tension force (T) and gravity $(m_B g)$. The tension forces must be equal and the magnitudes of the accelerations must be equal, since the cord connects the two blocks. (See Conceptual Question 3.) As block B accelerates downward, block A accelerates up the ramp. We added an *x-y* coordinate system separately to each free-body diagram, with the positive axes aligned with the direction of acceleration. The coordinate system for the right-hand block is rotated to coincide with the inclined plane. A rotated axis simplifies the analysis in ramp problems. This rotated axis requires resolving the gravity force into two components, one parallel and one perpendicular to the incline.

SOLVE We now apply Newton's second law to each block in the direction of acceleration to find the frictional force. Block A (with mass m_A) accelerates in the x direction (along the ramp), so

$$\sum F_x = T + (-m_A g \sin 35°) = m_A a.$$

Block B (with mass m_B) accelerates at the same rate in the y direction; hence,

$$\sum F_y = m_B g + (-T) = m_B a.$$

Both equations include the tension force, so we solve for that force in the second equation and substitute into the first. Our second equation then becomes

$$T = m_B g - m_B a.$$

Replacing the tension force in the first equation yields

$$(m_B g - m_B a) + (-m_A g \sin 35°) = m_A a.$$

Solving for the acceleration gives

$$m_B g + (-m_A g \sin 35°) = (m_A + m_B)a,$$

$$a = \frac{g(m_B - m_A \sin 35°)}{(m_A + m_B)} = \frac{(9.80 \text{ m/s}^2)((7.50 \text{ kg}) - (3.00 \text{ kg}) \sin 35°)}{((7.50 \text{ kg}) + (3.00 \text{ kg}))} = 5.39 \text{ m/s}^2.$$

The 3.00 kg block accelerates up the ramp at 5.39 m/s^2. The positive value of acceleration confirms the block's acceleration up the ramp.

REFLECT The value of acceleration is less than g, consistent with expectations. If the cord were cut, block B would accelerate with an acceleration equal to g. When block B is connected to block A, block B accelerates with an acceleration less than g. We say that block B has additional inertia when connected to block A.

What would have happened if we chose the acceleration direction incorrectly? We would have found a negative acceleration, indicating that the acceleration was down the incline. In this problem, the forces do not depend on the direction of motion, and a negative acceleration would not indicate an error. It does, however serve as a checkpoint for our calculation. A negative result with our choice of axes would cause us to be suspicious because the right-hand mass is larger and we expect it to accelerate downwards.

Practice Problem: Can you find the mass of block A when the system remains at rest? Will that mass simply be 7.50 kg? *Answer:* $m_1 = 13.1$ kg, no.

3: Frictional force on an accelerating block

Two blocks are connected to each other by a light cord passing over a small, frictionless pulley as shown in Figure 5.12. Block A has mass 5.00 kg and block B has mass 4.00 kg. If block B descends at a constant acceleration of 2.00 m/s^2 when set in motion, find the coefficient of kinetic friction between block A and the table.

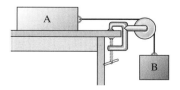

▲ **Figure 5.12** Problem 3.

Solution

SETUP To find the coefficient of kinetic friction, we will need to know the frictional force between block A and the table. We'll use Newton's laws to find the force. We begin with the free-body diagrams for the two blocks, shown in Figure 5.13.

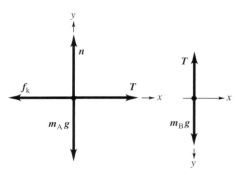

▲ **Figure 5.13** Problem 3 free-body diagram.

The forces are identified by their magnitudes. For block A, the forces are kinetic friction (f_k), gravity (m_Ag), the normal force (n), and the tension force (T). For block B, the forces are the tension force (T) and gravity (m_Bg). The tension forces are equal and the magnitudes of the accelerations are equal, as we have seen. As block B accelerates downward, block A accelerates to the right. To ensure that the acceleration of each block is in the positive direction, we added an x-y coordinate system separately to each free-body diagram, with the positive axes aligned with the direction of acceleration. All of the forces act along the coordinate axes, so we will not need to break the forces into components.

SOLVE We now apply Newton's second law to each block to find the frictional force. Block A (with mass m_A) accelerates in the x direction, so

$$\sum F_x = T + (-f_k) = m_A a.$$

Block B (with mass m_B) accelerates at the same rate in the y direction; thus,

$$\sum F_y = m_B g + (-T) = m_B a.$$

Both equations include the tension force, so we solve for that force in the second equation and substitute into the first. Our second equation is then

$$T = m_B g - m_B a = m_B(g - a).$$

Replacing the tension force in the first equation gives

$$m_B(g - a) + (-f_k) = m_A a,$$

Solving for the frictional force results in

$$f_k = m_B(g - a) - m_1 a = (4.00 \text{ kg})(9.80 \text{ m/s}^2 - 2.00 \text{ m/s}^2) - (5.00 \text{ kg})(2.00 \text{ m/s}^2) = 21.2 \text{ N}.$$

The frictional force is related to the coefficient of kinetic friction through the normal force. We find the normal force by examining the vertical components of the forces acting on block A. There is no acceleration in the vertical direction for block A, so we can apply the equilibrium condition to block A:

$$\sum F_y = n - m_A g = 0, \qquad n = m_A g.$$

Since there are no other vertical forces acting on block A, the normal force equals the weight. The kinetic frictional force is given by

$$f_k = \mu_k n,$$

which we can solve for μ_k:

$$\mu_k = \frac{f_k}{n} = \frac{f_k}{m_A g} = \frac{(21.2\ \text{N})}{(5.00\ \text{kg})(9.80\ \text{m/s}^2)} = 0.43.$$

We find that the coefficient of kinetic friction between the block and the table is 0.43.

REFLECT The coefficient of kinetic friction of 0.43 compares reasonably against values we've seen previously for smooth surfaces. Note that the tension is not equal to the weight of block B. That it is is a common misconception arising from examining the free-body diagram for block B without realizing that the block is accelerating. If we look at the rearranged second equation, this becomes clearer:

$$T = m_B(g - a).$$

The tension force is equal to the weight only when block B's acceleration is zero. Calculating the tension for this problem, we find a value of 31.2 N, 20% less than the weight of block B (39.2 N).

4: Constant-velocity motion of a crate up a *rough* inclined plane

A student pushes a crate up a rough inclined plane as shown in Figure 5.14. Find the magnitude of the horizontal force the student must apply for the crate to move up the incline at constant velocity. The crate has mass 15.0 kg, the incline is sloped at 30.0°, and the coefficient of kinetic friction between the crate and the incline is 0.600.

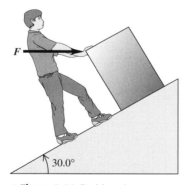

▲ **Figure 5.14** Problem 4.

Solution

SETUP There is no acceleration, so we will need to apply the equilibrium condition to the crate to find the applied force. We'll use a rotated coordinate system that coincides with the incline to simplify the analysis. The free-body diagram for the crate is shown in Figure 5.15.

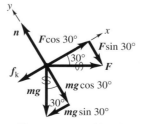

▲ **Figure 5.15** Problem 4
free-body diagram.

The forces are identified by their magnitudes: the kinetic-frictional force (f_k), gravity (mg), the normal force (n), and the applied force (F). The kinetic-frictional force opposes the motion up the incline and thus is directed down the incline. The rotated x-y coordinate system is indicated on the diagram. This rotated axis requires resolving the gravity and applied forces into components parallel and perpendicular to the incline, as shown in the diagram.

SOLVE We apply the equilibrium condition to the crate. In the x direction, along the incline, we have

$$\sum F_x = F\cos 30° + (-f_k) + (-mg\sin 30°) = 0.$$

We must apply the equilibrium condition in the y direction to find the normal force in order to quantify the frictional force. Thus,

$$\sum F_y = n + (-mg\cos 30°) + (-F\sin 30°) = 0.$$
$$n = mg\cos 30° + F\sin 30°.$$

The kinetic-frictional force is then

$$f_k = \mu_k n = \mu_k mg\cos 30° + \mu_k F\sin 30°.$$

We now substitute this result into the first equation:

$$F\cos 30° + (-\mu_k mg\cos 30° - \mu_k F\sin 30°) + (-mg\sin 30°) = 0.$$

Rearranging terms to solve for the applied force yields

$$F\cos 30° - \mu_k F\sin 30° = \mu_k mg\cos 30° + mg\sin 30°,$$
$$F(\cos 30° - \mu_k \sin 30°) = mg(\mu_k \cos 30° + \sin 30°),$$
$$F = \frac{mg(\mu_k \cos 30° + \sin 30°)}{(\cos 30° - \mu_k \sin 30°)} = \frac{(15.0 \text{ kg})(9.80 \text{ m/s}^2)((0.600)\cos 30° + \sin 30°)}{(\cos 30° - (0.600)\sin 30°)} = 265 \text{ N}.$$

The student must push with a horizontal force of 265 N to move the crate up the incline at constant velocity.

REFLECT A force of 265 N is roughly equivalent to the weight of a 27 kg object. Would it be easier to push the crate up the incline by pushing parallel to the incline? Yes, it would be easier pushing along the ramp. In this problem, the component of the applied force directed into the incline $(F\sin 30°)$ does nothing to move the crate up the ramp. In fact, this component increases the normal force and therefore the frictional force.

Practice Problem: If the student pushed solely along the incline, the problem would be simplified. What force would be necessary along the incline to maintain constant velocity of the crate? *Answer:* 150 N, 56% of the required horizontal force.

5: Spring force between two blocks

Two blocks are placed on a horizontal, frictionless surface and are attached to each other by a spring with force constant 4500 N/m. If the right-hand block is pulled with a force of 150.0 N, find the displacement of the spring as the blocks accelerate. The left-hand block has mass 5.00 kg, and the right-hand block has mass 3.00 kg.

Solution

SETUP The displacement of the spring is proportional to the spring force. We can find the spring force by applying Newton's second law to the blocks and then use Hooke's law to find the displacement. The first task is to sketch the situation, shown in Figure 5.16.

▲ **Figure 5.16** Problem 5.

Examining the sketch, we see that the two blocks interact through the spring. We draw free-body diagrams of the two blocks, shown in Figure 5.17.

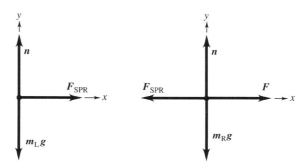

▲ **Figure 5.17** Problem 5 free-body diagram.

The forces are identified by their magnitudes: the spring force (F_{spr}), gravity (mg), the normal force (n), and the applied force (F). Knowledge of the vertical forces will not be necessary to solve this problem. The blocks are connected to each other; therefore, both accelerate at the same rate and are acted upon by the same magnitude of spring force. The diagrams include a common x-y coordinate axis with the positive x-axis in the direction of the acceleration. All of the forces act along the coordinate axes, so we will not need to break the forces into components.

SOLVE We now apply Newton's second law to the horizontal forces acting on each block to determine the spring force. For the right-hand block (with mass m_R),

$$\sum F_x = F + \left(-F_{spr}\right) = m_R a.$$

For the left-hand block (with mass m_L),

$$\sum F_x = F_{spr} = m_L a.$$

Examining these two equations, we find two unknowns: F_{spr} and a. We wish to find the spring force, so we rewrite the second equation in terms of acceleration:

$$a = \frac{F_{spr}}{m_L}.$$

Substituting for the acceleration in the first equation yields

$$F + (-F_{spr}) = m_R \frac{F_{spr}}{m_L},$$

$$F = F_{spr} + F_{spr} \frac{m_R}{m_L} = F_{spr}\left(1 + \frac{m_R}{m_L}\right),$$

$$F_{spr} = \frac{F}{\left(1 + \frac{m_R}{m_L}\right)} = F \frac{m_L}{m_R + m_L} = (150)\frac{(5.00 \text{ kg})}{(3.00 \text{ kg}) + (5.00 \text{ kg})} = 93.8 \text{ N}.$$

This gives us the magnitude of the spring force. The direction is opposite to the displacement. We can now use Hooke's law,

$$F_{spr} = -kx,$$

to solve for the displacement:

$$x = -\frac{F_{spr}}{k} = -\frac{(-93.8 \text{ N})}{4500 \text{ N/m}} = 0.0208 \text{ m} = 2.08 \text{ cm}.$$

The spring is displaced 2.08 cm when the blocks are pulled. We substituted a negative value for the spring force, as it opposes the displacement.

REFLECT We see that the spring force is less than the applied force in this problem. This is reasonable, as the spring must provide force to accelerate only the right-hand mass while the applied force must accelerate both masses.

For convenience, we did not include all forces in our analysis. By including all forces in free-body diagrams, however, we practice identifying forces and prepare for modified versions of the problem, such as adding friction or placing the blocks on an inclined plane.

Problem Summary

Chapters 4 and 5 have shown a variety of forces applied in diverse applications, but they share a common problem-solving foundation. For all problems, we

- Identified the general procedure for finding the solution.
- Sketched the situation when no figure was provided.
- Identified the forces acting on the objects.
- Drew free-body diagrams of the objects.
- Added appropriate coordinate systems to the free-body diagrams.
- Applied the equilibrium condition or Newton's second law (or both) to the objects in order to find relations between the forces, masses, and accelerations.
- Solved the equations through algebra and substitutions.
- Reflected on the results, thus checking for inconsistencies.

This problem-solving foundation can be applied to all problems involving forces. Following this procedure enables one to master problems involving Newton's laws.

6 CIRCULAR MOTION AND GRAVITATION

Summary

In this chapter, we will continue our study of dynamics, concentrating on circular motion. We will also delve deeper into the gravitational interaction by learning Newton's law of gravitation and gaining a better understanding of the concept of weight. We'll see how to combine the concepts of circular motion and gravitation to explain the orbits of satellites and planets.

Objectives

- Learn to apply Newton's laws to uniform circular motion.
- Learn Newton's law of gravitation and apply it to pairs of masses.
- Learn the general definition of weight.
- Understand how satellites orbit astronomical bodies.

Concepts and Equations

Term	Description
Period	The period of rotation is the time an object takes to make one revolution. The period is denoted by T.
Expressions for Rotational Motion	Useful expressions for objects in uniform circular motion include $$v = \frac{2\pi R}{T} \qquad \text{(velocity in terms of radius and period)},$$ $$a_{\text{rad}} = \frac{4\pi^2 R}{T^2} \qquad \text{(radial component of acceleration)},$$ $$F_{\text{net}} = m\frac{v^2}{R} \qquad \text{(net radial force acting on an object in circular motion)},$$ where R is the radius of the circle and m is the object's mass.
Newton's Law of Gravitation	Newton's law of gravitation states that the magnitude of the force between two particles with masses m_1 and m_2, separated by a distance r, is given by $$F_{\text{g}} = G\frac{m_1 m_2}{r^2},$$ where G denotes the gravitational constant and is equal to $6.67 \times 10^{-11}\ \text{N} \cdot \text{m}^2/\text{kg}^2$. The gravitational force is always attractive, pointing along the line that separates the objects.
Weight	The weight of an object is the total gravitational force exerted on the object by all other objects in the universe. Near the surface of the earth, an object's weight is very nearly equal to the gravitational force of the earth on the object alone.
Satellite Motion	For a satellite moving in a circular orbit, the gravitational attraction between the satellite and the astronomical body provides the centripetal acceleration. The velocity v and period T for a satellite orbiting at radius r is given by $$v = \sqrt{\frac{GM}{r}},$$ $$T = \frac{2\pi r^{3/2}}{\sqrt{GM}},$$ where M is the mass of the astronomical body.

Conceptual Questions

1: Turning in a car

As you make a right turn in your car, what pushes you against the car door?

Solution

SET UP AND SOLVE As the car turns, your body tends to continue moving in a straight line; therefore, you push up against the car door. So you can say that *no force* pushes you against the door and your body tends to maintain a constant velocity due to Newton's first law. Once the car door comes in contact with you, it pushes you in the direction of the turn, accelerating you to the right.

REFLECT From within the car, you may wonder what force pushes you to the side. However, since the car is turning to the right, it is accelerating, and the car is not an inertial reference frame. Therefore, we cannot apply Newton's laws inside the car. If we consider how the situation would appear to someone outside of the car (in an inertial reference frame), we can apply Newton's laws. You would then appear to continue in a straight line while the car moves to the right. You may want to review Conceptual Question 2 in Chapter 4 for a similar case.

2: Free-body diagram of a car on a hill

Draw a free-body diagram of a car going over the top of a round hill at a constant speed. Is there a net force acting on the car if it is moving at constant speed?

Solution

SET UP AND SOLVE Two forces act on the car at the top of the hill: the normal force due to the road and gravity. The normal force is upward and gravity is downward. Figure 6.1 shows the free-body diagram.

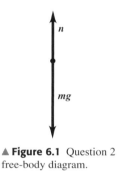

n

mg

▲ **Figure 6.1** Question 2 free-body diagram.

The force vectors in the free-body diagram are not drawn with equal length; the gravitational force is larger than the normal force. This is because there is a net force acting on the car; the car's velocity is changing direction, so the car has a centripetal acceleration. The centripetal acceleration is downward, so the net force must be downward.

There is a net force acting on the car even though the car is moving at constant speed.

REFLECT Constant speed does not necessarily imply constant velocity. You must carefully interpret problems involving circular motion and constant speed.

3: Is the earth falling?

There is a net gravitational force between the earth and the sun, so why doesn't the earth fall into the sun?

Solution

SET UP AND SOLVE The earth is constantly falling towards the sun, but the earth doesn't get closer to the sun, since the sun's surface curves away beneath the earth. Recall projectile motion from Chapter 3. If we launch an object parallel to the ground, it follows a parabolic path. If we give the object a larger initial velocity, then the object moves farther away from the launch site as it falls. The earth is round, so as the object moves farther away, the object will have a greater distance to fall. At a sufficiently high launch velocity, the object will make a complete revolution and not land on the ground. This is the same physical situation as the earth revolving around the sun. If the earth had a smaller velocity, it would fall into the sun.

REFLECT This result may seem a bit odd, but it is indeed accurate. It also shows how understanding one physical phenomenon helps us understand other phenomena: Our experience with projectile motion helped us interpret the motion of the earth around the sun.

4: Does the earth maintain constant speed?

In the last question, we saw that the earth is constantly falling. Does it maintain constant speed if it is falling?

Solution

SET UP AND SOLVE There is a net gravitational force acting on the earth due to the sun. The direction of the net force is towards the sun; however, the earth's velocity is perpendicular to the direction of force. The gravitational force can change only the direction of the earth's velocity around the sun and not the magnitude. Thus, the earth maintains constant speed as it orbits the sun.

REFLECT We've come to know that a net force causes acceleration—a change in velocity. In the previous chapters, the magnitude of an object's velocity often changed when the object was acted upon by a net force. This chapter examines additional consequences of the influence of forces.

Problems

1: Coefficient of friction on a banked curve

A circular section of road with a radius 150 m is banked at an angle of 12°. What should be the minimum coefficient of friction between the tires and the road if the roadway is designed for a speed of 25 m/s?

Solution

SET UP Figure 6.2 is a free-body diagram of the car tire on the road. The figure shows the three forces acting on the tire: the normal force due to the road (n), static friction with the road (f), and gravity (mg). For the tire not to slip, the vertical forces must be in equilibrium and a net horizontal force must act towards the center of the circular section of road.

▲ **Figure 6.2** Problem 1 free-body diagram

We have added an x-y coordinate system to the figure because the forces act in two dimensions. Note that we aligned the axes horizontally and vertically to coincide with the net forces. We've also resolved the normal and frictional forces into their x and y components.

SOLVE We apply the force conditions to the tire, writing separate equations for the x and y components. In the vertical direction, there is no net force:

$$\sum F_y = 0, \qquad n\cos 12° - f\sin 12° - mg = 0.$$

In the horizontal direction, we use Newton's second law with centripetal acceleration:

$$\sum F_x = ma_\text{rad}, \qquad n\sin 12° + f\cos 12° = m\frac{v^2}{r}.$$

Note that the downward components are negative, consistent with our coordinate system. The static frictional force can be replaced with μn in our two equations:

$$n\cos 12° - \mu n\sin 12° - mg = 0,$$
$$n\sin 12° + \mu n\cos 12° = m\frac{v^2}{r}.$$

We now rewrite the first equation in terms of n and substitute into the second:

$$n = \frac{mg}{\cos 12° - \mu\sin 12°},$$

$$\frac{mg}{\cos 12° - \mu\sin 12°}\sin 12° + \mu\frac{mg}{\cos 12° - \mu\sin 12°}\cos 12° = m\frac{v^2}{r}.$$

The mass cancels and we can solve for μ:

$$\mu = \frac{v^2\cos 12° - rg\sin 12°}{v^2\sin 12° + rg\cos 12°} = \frac{(25\text{ m/s})^2\cos 12° - (150\text{ m})(9.8\text{ m/s}^2)\sin 12°}{(25\text{ m/s})^2\sin 12° + (150\text{ m})(9.8\text{ m/s}^2)\cos 12°} = 0.20.$$

The minimum coefficient of static friction between the tire and road is 0.20.

REFLECT The technique for solving this problem is similar to those set forth in the previous chapter. The differences in this problem were the inclusion of centripetal acceleration and choosing axes that corresponded to our knowledge of the net forces.

Practice Problem: What speed would require no frictional force? *Answer:* $v = 18$ m/s.

2: Normal force on a roller coaster

A roller coaster has a vertical loop of radius 45 m. If the roller coaster operates at a constant speed of 35 m/s while in the loop, what normal force does the seat exert on a 75 kg passenger at the top of the loop? The roller coaster is upside down at the top of the loop.

Solution

SET UP To find the normal force, we'll sum the forces acting on the passenger. Figure 6.3 is a free-body diagram of the passenger on the roller coaster. The figure shows the two forces acting on the passenger: the normal force due to the seat (n) and gravity (mg). At the top of the loop, the net force is downward and the person is accelerating towards the center. We have added an x-y coordinate system to the figure; positive forces are downward.

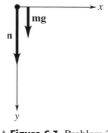

▲ **Figure 6.3** Problem 2
free-body diagram

SOLVE The net force on the passenger is downward. Newton's second law with centripetal acceleration gives

$$\sum F_y = ma_{\text{rad}}, \qquad n + mg = m\frac{v^2}{r}.$$

Solving for the normal force yields

$$n = m\frac{v^2}{r} - mg = (75 \text{ kg})\frac{(35 \text{ m/s})^2}{(45 \text{ m})} - (75 \text{ kg})(9.8 \text{ m/s}^2) = 1300 \text{ N},$$

The seat exerts a force of 1,300 N on the passenger.

REFLECT We see that the seat exerts a force nearly twice the passenger's weight. A seat belt would not be needed to prevent a fall from the roller coaster at the top of the loop. Amusement park rides get their reputation for excitement from their ability to rapidly change the magnitude and direction of force applied to passengers. The normal force of the seat on the passenger is even larger at the bottom of the loop.

Practice Problem: What normal force does the seat provide at the bottom of the loop? *Answer:* $N = 2,800$ N.

3: Gravitational force due to three masses

Three masses are arranged as shown in Figure 6.4. Find the net force acting on the top mass (*A*). Each mass is 5.00 kg.

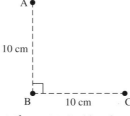

▲ **Figure 6.4** Problem 3.

Solution

SET UP The net force on *A* is the sum of the forces between *A* and *B* and between *A* and *C*. A free-body diagram illustrating these two forces is shown in Figure 6.5. We'll need to add the two forces by using their components. Newton's law of gravitation gives the magnitude of the forces. We'll use the coordinate axes provided in the figure.

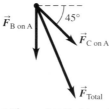

▲ **Figure 6.5** Problem 3 free-body diagram.

SOLVE To apply Newton's law of gravity, we need to know the distances between the masses. By summing the squares of the sides of the triangle and taking the square root, we find that the distance between A and B is 10.0 cm and the distance between A and C is 14.1 cm. The force of B on A is

$$F_{B \text{ on } A} = \frac{Gm_B m_A}{r_{BA}^2} = \frac{(6.67 \times 10^{-11} \text{ N} \cdot \text{m}^2/\text{kg}^2)(5.00 \text{ kg})(5.00 \text{ kg})}{(0.100 \text{ cm})^2} = 1.67 \times 10^{-7} \text{ N}.$$

The force of C on A is

$$F_{C \text{ on } A} = \frac{Gm_C m_A}{r_{CA}^2} = \frac{(6.67 \times 10^{-11} \text{ N} \cdot \text{m}^2/\text{kg}^2)(5.00 \text{ kg})(5.00 \text{ kg})}{(0.141 \text{ cm})^2} = 8.39 \times 10^{-8} \text{ N}.$$

With the magnitudes of the forces determined, we simply add the two vectors together, using their components. The force of C on A has the only x component:

$$\sum F_x = F_{C \text{ on } A} \sin 45° = 5.93 \times 10^{-8} \text{ N}.$$

The 45° angle results because the masses are arranged as an isosceles triangle. Both forces have y components:

$$\sum F_y = -F_{B \text{ on } A} - F_{C \text{ on } A} \sin 45° = -2.26 \times 10^{-7} \text{ N}.$$

The negative result indicates that the y component points downward. We find the magnitude of the net force by combining the components:

$$F = \sqrt{F_x^2 + F_y^2} = 2.34 \times 10^{-7} \text{ N}.$$

The direction of the net force is found by using the tangent. We want to specify the angle ϕ with respect to the x axis:

$$\phi = \tan^{-1}\frac{F_y}{F_x} = \tan^{-1}\frac{(-2.26 \times 10^{-7} \text{ N})}{(5.93 \times 10^{-8} \text{ N})} = -75.3°.$$

The net force on A has magnitude 2.34×10^{-7} N and points 75.3° below the positive x axis.

REFLECT We see that the gravitational force between masses is very small. To have an appreciable gravitational force, we need to have at least one large mass, such as the earth. Also, we can clearly see that Newton's third law is valid: Reversing subscripts in the first two equations would result in a force of the same magnitude, but opposite in direction.

4: Orbit of a weather satellite

Imagine that you are designing a new weather satellite. The goal is to have the satellite orbit the earth in a circular orbit every 6 hours. At what distance above the earth's surface should the satellite be placed to obtain the correct period?

Solution

SET UP The force acting on the satellite is the gravitational force between the satellite and the earth. The satellite follows a circular orbit, so it has radial acceleration towards the center of the earth. We will use this information to find the velocity of the satellite that can be used to find the period of rotation. We solve for the distance by setting the period to 6 hours.

SOLVE Newton's law of gravitation gives the force on the satellite due to the earth:

$$F_g = \frac{Gmm_E}{r^2}.$$

Here, m is the mass of the satellite, m_E is the mass of the earth, and r is the distance from the center of the earth to the satellite. Newton's second law gives the net force on the satellite (the acceleration is v^2/r):

$$\sum F = ma_{\text{rad}}, \qquad \sum F = F_g = \frac{Gmm_E}{r^2} = m\frac{v^2}{r}.$$

Solving for v, we find that

$$v = \sqrt{\frac{Gm_E}{r}}.$$

We can also write the velocity in terms of the distance the satellite travels $(2\pi r)$ in one period (T):

$$v = \frac{2\pi r}{T}.$$

To find the radius, we equate the last two equations and solve for r, obtaining

$$r = \sqrt[3]{\frac{Gm_E T^2}{4\pi^2}} = \sqrt[3]{\frac{(6.67 \times 10^{-11}\ \text{N} \cdot \text{m}^2/\text{kg}^2)(5.98 \times 10^{24}\ \text{kg})(2.16 \times 10^4\ \text{s})^2}{4\pi^2}} = 1.68 \times 10^7\ \text{m},$$

where we replaced the 6 hour period with the equivalent 21,600 s. The satellite should be placed in an orbit of radius 16,800 km. Subtracting the radius of the earth (6380 km) from this radius, we find that the satellite should be placed 10,400 km above the earth's surface to achieve a 6 hour orbital period.

REFLECT We could have avoided our derivation and used Equation 6.10 from the textbook to arrive at the solution directly. However, the review presented here helps remind us how to find the solution without searching through the book.

Practice Problem: What does the free-body diagram for the satellite look like? *Answer:* A single vector.

WORK AND ENERGY

Summary

We'll introduce a new concept in this chapter: *energy*. Energy is one of the most important concepts in physics, due to the principle of conservation of energy. We'll concentrate on mechanical energy and work, gaining experience with common forms of energy, including kinetic energy (the energy of motion) and potential energy (a form of energy storage). By the end of the chapter, we'll be able to apply energy concepts to the analysis of problems, and we'll be prepared to include additional forms of energy that we'll encounter in later chapters.

Objectives

- Learn the definition of work and how to calculate the work done by a constant force.
- Learn the definition of kinetic energy, gravitational potential energy, and elastic potential energy.
- Learn the work–energy theorem and the principle of conservation of energy and apply them to problems.
- Define and calculate power for objects undergoing work.

Concepts and Equations

Terms	Description
Energy	Energy is a quantity that is capable of producing change, that comes in a variety of forms, and that can be exchanged and transformed in a system. Energy is measured in joules (J); 1 joule is equal to 1 newton-meter.
System	A system is a collection of one or more objects that can interact, move, and deform. An isolated system has no interactions with its surroundings.
Work	A constant force acting on and displacing an object does work. For a constant force $\vec{F}$ with magnitude F causing a displacement $\vec{s}$ with magnitude s at an angle ϕ with respect to the force, the work done by the force on the object is $$W = F_{\parallel}s = (F\cos\phi)s,$$ where $F_{\parallel}$ is the component of the force parallel to the direction of displacement.
Mechanical Energy	Mechanical energy is the energy associated with motion, position, and deformation.
Kinetic Energy	Kinetic energy is the energy of motion for particles with mass. Kinetic energy is denoted by K. A particle of mass m and velocity v has kinetic energy $$K = \tfrac{1}{2}mv^2.$$
Potential Energy	Potential energy is stored energy that has the potential to do work. There are several varieties of potential energy, including gravitational potential energy and elastic potential energy.
Gravitational Potential Energy	Gravitational potential energy is the potential energy associated with the position of an object relative to earth. For an object of mass m at a vertical distance y above the origin in a uniform gravitational field g, the gravitational potential energy of the system consisting of the object and the earth is $$U_{\text{grav}} = mgy.$$ Gravitational potential energy does not depend upon the location of the origin; only *differences* in gravitational potential energy are significant.
Elastic Potential Energy	Elastic potential energy is the potential energy associated with the position of a spring that obeys Hooke's law. For a spring of force constant k stretched or compressed a distance x from equilibrium, the elastic potential energy is $$U_{\text{el}} = \tfrac{1}{2}kx^2.$$
Work–energy Theorem	The work–energy theorem states that the work done by a net external force on a particle is equal to the change in kinetic energy of the particle: $$W_{\text{total}} = K_f - K_i = \Delta K.$$
Conservation of Energy	When only conservative forces act on an object, the total mechanical energy is constant and is given by $$K_i + U_i = K_f + U_f,$$ where U includes both gravitational and elastic potential energies.
Power	Power is the quantity of work done per unit time. Average power is defined as $$P_{\text{av}} = \frac{\Delta W}{\Delta t},$$ where ΔW is the quantity of work performed in time interval Δt.

Conceptual Questions

1: Work done by the normal force

How much work does the normal force do on a box sliding across the floor?

Solution

SET UP AND SOLVE Work is produced when a force moves an object in the direction the object is displaced. As a box slides across the floor, the normal force is perpendicular to its motion. Therefore, the normal force does no work on the sliding box.

REFLECT Work has a strict definition in physics. The normal force does prevent the box from falling into the floor, but it does no work thereby. You also do no work when you support your backpack as you walk down the hall, even though your arm tires.

2: Which force does the most work?

Rank the four situations shown in Figure 7.1 from most to least work done by the force. The displacement is the same in each case.

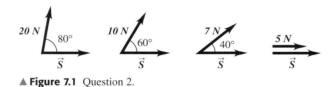

▲ **Figure 7.1** Question 2.

Solution

SET UP AND SOLVE Work is displacement times the component of a force parallel to the displacement. All four situations have the same displacement, so we rank the components of the force parallel to the displacement. The parallel components of force are the magnitudes of the forces times the cosine of the angle between the force and the displacement vectors.

We find that the parallel components for the four situations are, in order, 3.47 N, 4.00 N, 5.36 N, and 5.00 N. Therefore, the ranking from greatest to least amount of work is (c), (d), (b), and (a).

REFLECT We see that larger forces do not necessarily produce more work. Work depends on both the magnitude of the force and the direction of the force with respect to the displacement. In this example, the 20 N force produces much less work than the 5 N force.

3: Multiple routes to bottom of a hill

Figure 7.2 shows four different routes to the bottom of a hill, all starting from the same initial height. If you and three of your friends slide down the four routes, how do the four speeds at the bottom of the hill compare? Each of the paths is frictionless and everyone starts from rest.

▲ **Figure 7.2** Question 3.

Solution

SET UP AND SOLVE Gravitational potential energy depends only on *changes* in height. Therefore, the change in gravitational potential energy is the same for all four routes, because all of them have the same change in height. The kinetic energy at the bottom of the hill will be the same for all four friends; therefore, the speeds of all four friends will be the same at the bottom.

REFLECT If the four speeds are the same at the bottom, what quantity differs? You can see that the four routes have different lengths and are shaped differently. If you compare path 2 with path 3, you see that path 3 has a steep initial drop-off, while path 2 has a shallower initial drop-off. The friend on path 3 will initially accelerate faster than the friend on path 2, have a larger speed throughout, and arrive at the bottom first. Therefore, the time required to reach the bottom differs for the different paths.

Problems

1: Work done in pushing a crate up an inclined plane

A student pushes a crate 3.50 m up a rough inclined plane with a constant horizontal force of 225 N starting from rest, as shown in Figure 7.3. Find the work done by the student, the work done by friction, the change in gravitational potential energy, and the change in kinetic energy. What is the sum of these energies? The crate has mass 15.0 kg, the incline is sloped at 30.0°, and the coefficient of kinetic friction between the crate and the incline is 0.400.

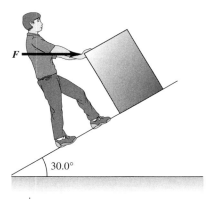

▲ **Figure 7.3** Problem 1.

Solution

SET UP We'll begin with a free-body diagram and calculate the separate quantities by using definitions of work and energy. Figure 7.4 shows the free-body diagram with a rotated coordinate system that coincides with the incline to simplify the analysis. The forces are identified by their magnitudes: the kinetic-frictional force (f_k), gravity (mg), the normal force (n), and the applied force (F). The kinetic-frictional force opposes the motion up the incline and thus is directed down the incline. Gravity and the applied force are resolved into components parallel and perpendicular to the incline.

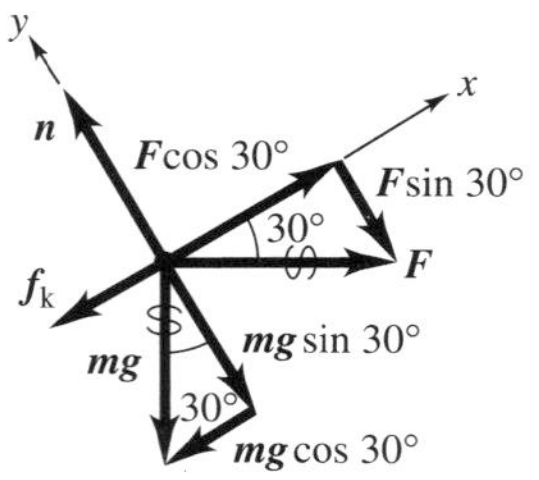

▲ **Figure 7.4** Problem 1
free-body diagram.

SOLVE The work done by the student pushing the crate is

$$W = F_\parallel s = (F\cos 30°)s = (225\text{ kg})(\cos 30°)(3.50\text{ m}) = 682\text{ J}.$$

To find the work done by friction, we need to know the frictional force. We apply the equilibrium condition in the y-direction to find the normal force in order to quantify the frictional force:

$$\sum F_y = n + (-mg\cos 30°) + (-F\sin 30°) = 0,$$
$$n = mg\cos 30° + F\sin 30° = (15.0\text{ kg})(9.8\text{ m/s}^2)(\cos 30°) + (225\text{ N})(\sin 30°) = 240\text{ N}.$$

The kinetic frictional force is then

$$f_k = \mu_k n = (0.400)(240\text{ N}) = 95.9\text{ N}.$$

The work done by friction is

$$W = F_\parallel s = (f_k)s = (95.9\text{ N})(3.50\text{ m}) = 336\text{ J}.$$

The change in gravitational potential energy is given by

$$\Delta U_{\text{grav}} = U_f - U_i = mgy_f - mgy_i.$$

The initial position is zero, so y_i is zero. The final position is found by trigonometry. We have

$$\Delta U_{\text{grav}} = mgy_f = (15.0\text{ kg})(9.8\text{ m/s}^2)(3.5\text{ m})\sin 30° = 257\text{ J}.$$

To find the change in kinetic energy, we need the initial and final velocities. The initial velocity is zero. The final velocity is found with Newton's second law and kinematics. In the x-direction, along the incline, Newton's second law gives

$$\sum F_x = F\cos 30° + (-f_k) + (-mg\sin 30°) = ma.$$

The acceleration of the box is then

$$a = \frac{F\cos 30° + (-f_k) + (-mg\sin 30°)}{m}$$
$$= \frac{(225\text{ N})\cos 30° - (95.9\text{ N}) - (15.0\text{ kg})(9.8\text{ m/s}^2)\sin 30°}{(15.0\text{ kg})}$$
$$= 1.70\text{ m/s}^2.$$

Constant-acceleration kinematics gives the final velocity:

$$v^2 = v_{x0}^2 + 2a_x(x - x_0),$$
$$v = \sqrt{0 + 2(1.70\text{ m/s}^2)(3.50\text{ m})} = 3.45\text{ m/s}.$$

The change in kinetic energy is then

$$\Delta K = K_f - K_i = \tfrac{1}{2}mv^2 - 0 = \tfrac{1}{2}(15.0 \text{ kg})(3.45 \text{ m/s})^2 = 89.3 \text{ J}.$$

Summarizing, we found that the student did 682 J of work on the crate and friction removed 336 J of work, leaving 346 J available for changing the kinetic and potential energies. The gravitational potential energy increased by 257 J and the kinetic energy increased by 89 J, a total of 346 J. We have accounted for all of the energy in the problem.

REFLECT This problem involves a thorough investigation of work, energy, and the principle of conservation of energy. It illustrates how energy is transformed among four varieties.

2: Professor landing on spring platform

Your professor, with mass 60.0 kg, falls from a height of 2.50 m onto a platform mounted on a spring. As the spring compresses, she compresses the spring a maximum distance of 0.240 m. What is the force constant of the spring? Assume that the spring and platform have negligible mass.

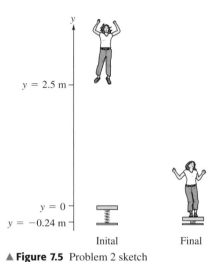

▲ **Figure 7.5** Problem 2 sketch

Solution

SET UP Figure 7.5 shows a sketch of the situation. This is a conservative system, as the only forces acting on the professor are gravity and the spring force. We'll set the origin at the top of the uncompressed platform. Initially, the professor has only U_{grav}, since her velocity is zero $(K = 0)$ and the spring is uncompressed $(U_{\text{el}} = 0)$. As she falls to the top of the platform, her kinetic energy increases and gravitational potential energy decreases. As she starts to compress the spring, she slows down as energy is transformed to the spring's elastic potential energy. At maximum compression, she comes to a momentary stop $(K = 0)$. At this point, she is below the origin and has negative gravitational potential energy. We'll use energy conservation to solve for the spring's force constant.

SOLVE Energy conservation relates the initial and final energies:

$$K_i + U_i = K_f + U_f.$$

Initially, there is only U_{grav}. At the maximum compression, there are two potential-energy terms, U_{grav} and U_{el}, and

$$U_{\text{grav},i} = U_{\text{grav},f} + U_{\text{el},f}.$$

Replacing with the expressions for the energies yields

$$mgy_i = mgy_f + \tfrac{1}{2}kx^2.$$

The initial height is 2.50 m, and the final height and the compression is -0.240 m. Solving for k gives

$$k = \frac{2mg(y_i - y_f)}{x^2} = \frac{2(60.0\text{ kg})(9.80\text{ m/s}^2)(2.50\text{ m} - (-0.240\text{ m}))}{(0.240\text{ m})^2} = 55{,}900\text{ N/m}$$

The force constant of the spring is 55,900 N/m.

REFLECT Our choice of origin gave a negative y_f, but only *differences* in gravitational potential energies influence the result. This is a problem that we could not have solved with our force techniques, as the force of the spring varies with position.

3: Designing a bungee jump

You are entering the bungee-jumping business and must design the bungee cord. The jump will be from a bridge that is 100.0 m above a river. The design calls for 2.00 seconds of free fall before the cord begins to slow the fall, and the person just touches the water after jumping. Find the force constant and length of the bungee cord for a 100.0 kg person.

Solution

SET UP Figure 7.6 shows a sketch of the situation, with the coordinate origin at the river. The forces acting on the jumper are gravity and the spring force. There is no mechanical work done on the system, so energy is conserved. On the bridge, the jumper has gravitational potential energy. After he jumps, the energy transforms to kinetic and elastic potential energies. At the river, the jumper momentarily stops and all the energy has transformed into elastic potential energy.

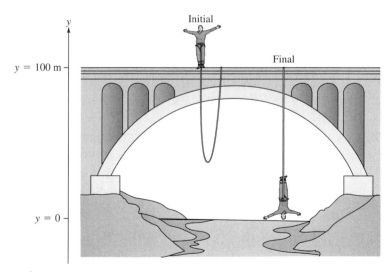

▲ **Figure 7.6** Problem 3.

We'll need to recall our kinematics for freely falling objects to find the length of the bungee cord. The length of the cord is found by determining its length when it becomes taut. We know that the cord becomes taut after 2.00 s, so we can use free-fall kinematics to solve for the distance the person falls in

2.00 s, which is equal to the length of the bungee cord. We'll ignore air resistance and any friction in the cord.

SOLVE Energy conservation relates the initial and final energies:

$$K_i + U_i = K_f + U_f.$$

Initially, there is only U_{grav} at the bridge. At the river, there is only U_{el}. Thus,

$$U_{grav,i} = U_{el,f}.$$

Replacing with the expressions for the energies gives

$$mgy_i = \tfrac{1}{2}kx^2,$$

where m is the mass of the jumper, y_i is height of the bridge, x is the bungee cord stretch length, and k is the force constant of the bungee cord. We need the amount of stretch in the bungee cord, so we first find the position where the bungee cord becomes taut, using constant-acceleration kinematics for freely falling objects:

$$y_{taut} = y_0 + v_{0y}t + \tfrac{1}{2}a_y t^2.$$

Here, v_{0y} is zero as the person starts from rest, $y_0 = 100$ m, $a_y = -g$, and t is 2.00 s. Solving for the position y_{taut} where the bungee cord becomes taut yields

$$y_{taut} = 100 \text{ m} + \tfrac{1}{2}(-9.8 \text{ m/s}^2)(2.00 \text{ s})^2 = 80.4 \text{ m}.$$

Thus, $y_{taut} = 80.4$ m is the vertical position above the river where the bungee cord becomes taut. The starting point is $y_0 = 100$ m, so the length of the bungee cord is the difference between y_0 and y_{taut}: $100 \text{ m} - 80.4 \text{ m} = 19.6$ m. The stretch length is how much the cord is stretched from its original length, or 80.4 m in this case (i.e., the distance from where the bungee cord becomes taut to the river). Substituting into our energy expression to solve for the force constant, we get

$$k = \frac{2mg(y_0)}{x^2} = \frac{2(100.0 \text{ kg})(9.8 \text{ m/s}^2)(100.0 \text{ m})}{(80.4 \text{ m})^2} = 30.3 \text{ N/m}.$$

You will need a bungee cord that is 19.6 m long with a 30.3 N/m force constant.

REFLECT The spring constant was found to be relatively small, indicating that the person will be slowed gently. What will happen to a person with a mass of less than 100 kg? What about a person with mass greater than 100 kg? The lighter person has less initial energy and so stops above the river. The heavier person has more initial energy and so stops under the surface of the river (so the cord should be changed!).

4: Toy car loop-the-loop

A toy car is released from a spring launcher onto a horizontal track that leads to a vertical loop-the-loop as shown in Figure 7.7. What is the minimum compression needed for the launcher so that, when released, the car remains on the track throughout the loop? The mass of the car is 10.0 g, the force constant of the launcher is 20.0 N/m, the loop has a radius of 20.0 cm, and you may assume that the car moves along the track without friction.

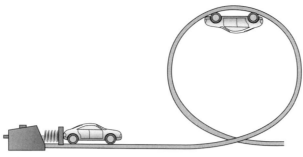

▲ **Figure 7.7** Problem 4.

Solution

SET UP The forces acting on the car are gravity, the spring force of the launcher while the car is in contact with it, and the normal force of the ground or loop. There is no mechanical work done on the system, so energy is conserved.

For the car to remain in contact with the track at the top of the loop, it must have sufficient velocity to maintain centripetal force, as we saw in Chapter 6. A free-body diagram of the car at the top of the loop is shown in Figure 7.8.

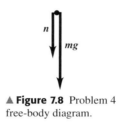

▲ **Figure 7.8** Problem 4 free-body diagram.

We place the origin at the ground level. Our initial point (*i*) will be when the car is at rest and the launcher is compressed, storing all the energy in the spring. After the car is released, the elastic potential energy transforms into kinetic energy and then into a combination of kinetic and gravitational potential energy when it enters the loop. The final point (*f*) will be at the top of the loop, where there are both kinetic and gravitational potential energies. We'll solve for the spring's compression by using energy conservation.

SOLVE Energy conservation relates the initial and final energies:

$$K_i + U_i = K_f + U_f.$$

Initially, there is only U_{el} stored in the spring. At the top of the loop, both K and U_{grav} are stored, and

$$U_{el,i} = U_{grav,f} + K_f.$$

Replacing with the expressions for the energies gives

$$\tfrac{1}{2}kx^2 = mgy_f + \tfrac{1}{2}mv^2,$$

where x is the spring compression, k is the force constant of the spring, m is the mass of the car, y_f is twice the radius of the loop (the height at the top of the loop), and v is the speed of the car at the top of the loop. We find the velocity at the top of the loop by applying Newton's second law. To find the minimum compression of the spring, we need the minimum velocity at the top of the loop. The minimum velocity corresponds to the minimum force on the car at the top; therefore, the only force acting on the car at the top is gravity, so

$$\sum F_y = mg = ma_{\text{rad}} = \frac{mv^2}{r}.$$

Solving for v yields the velocity at the top of the loop:

$$v = \sqrt{gr} = \sqrt{(9.80 \text{ m/s}^2)(0.200 \text{ m})} = 1.40 \text{ m/s}.$$

Combining the results and solving for the displacement of the spring gives

$$x = \sqrt{\frac{m(4gr + v^2)}{k}} = \sqrt{\frac{0.0100 \text{ kg}(4(9.80 \text{ m/s}^2)(0.200 \text{ m}) + (1.40 \text{ m/s})^2)}{20.0 \text{ N/m}}} = 0.0700 \text{ m}.$$

The minimum spring compression necessary to keep the car on the track throughout the loop is 7.00 cm.

REFLECT We found the minimum compression of the spring. Additional compression would have resulted in greater total energy after the car is launched, which would also keep the car on the track.

This problem illustrates how we'll sometimes combine previous knowledge (e.g., forces) with our current topics. As we progress through the text, we will add to our knowledge base and not exchange one concept for another.

Problem Summary

This chapter has augmented our knowledge of forces and kinematics with our newly formed knowledge of energy analysis. Energy analysis shares many of the problem-solving principles we have encountered. In these problems, we

- Identified the general procedure for finding the solution.
- Sketched the situation when no figure was provided.
- Identified the energies involved and the forces acting in the system.
- Applied energy principles, including conservation of energy (when possible).
- Drew free-body diagrams for the objects when appropriate.
- Applied Newton's laws when appropriate.
- Solved the equations through algebra, including substitutions.
- Reflected on the results, thus checking for inconsistencies.

This problem-solving foundation can be applied to all problems, including those which could be solved by force analysis. Following this procedure enables one to master many physics problems.

8   MOMENTUM

Summary

We'll introduce two new concepts in this chapter, *momentum* and *impulse*, which will serve as our third major analysis technique in mechanics. Like energy analysis, momentum and impulse analysis will expand our problem-solving repertoire and allow us to tackle collision problems that would be challenging with Newton's laws. Also, like energy, momentum is a conserved quantity that has important consequences throughout physics. We'll be able to apply momentum and impulse analysis to a wide variety of problems by the end of the chapter, when, combining it with force and energy analyses, we will be able to apply a powerful set of tools to investigate many natural phenomena.

Objectives

- Learn the definition of momentum and impulse.
- Restate Newton's law in terms of momentum.
- Apply conservation of momentum to collisions.
- Use the impulse to find the average force during a collision.
- Know how to identify elastic, inelastic, and totally inelastic collisions.
- Find the center of mass of a system.

Concepts and Equations

Term	Description
Momentum	The momentum $\vec{p}$ of a particle of mass m moving with velocity $\vec{v}$ is defined as $$\vec{p} = m\vec{v}.$$ Newton's second law can be written $$\sum \vec{F} = \lim_{\Delta t \to \infty} \frac{\Delta \vec{p}}{\Delta t}.$$ The total momentum $\vec{P}$ of a system of particles is the vector sum of the individual momenta: $$\vec{P} = \vec{p}_A + \vec{p}_B + \cdots = m_A\vec{v}_A + m_B\vec{v}_B + \cdots.$$
Conservation of Momentum	The total momentum of a system is constant when the vector sum of the external forces on the system is zero. For an isolated system, $$\vec{P}_i = \vec{P}_f.$$ Components of momentum are conserved separately.
Elastic Collision	In an elastic collision between two objects, the kinetic energy is conserved and the initial and final relative velocities have equal magnitude, but opposite direction.
Inelastic Collision	In an inelastic collision between two objects, the final kinetic energy is less than the initial kinetic energy. If the two objects have the same final velocity, the collision is completely inelastic.
Impulse	The impulse of a constant force, denoted $\vec{J}$, is the force multiplied by the time interval over which the force acts: $$\vec{J} = \vec{F}(t_f - t_i) = \vec{F}\Delta t.$$
Impulse–momentum Theory	The impulse–momentum theory states that the change in an object's momentum during a time interval equals the impulse of the total force acting on the object, or $$\Delta \vec{p} = \vec{F}\Delta t = \vec{J}.$$
Center of Mass	The coordinates of the center of mass of a system of particles is given by $$x_{cm} = \frac{m_A x_A + m_B x_B + \cdots}{m_A + m_B + \cdots},$$ $$y_{cm} = \frac{m_A y_A + m_B y_B + \cdots}{m_A + m_B + \cdots}.$$ The total momentum of the system equals the total mass multiplied by the velocity of the center of mass: $$\vec{P} = M\vec{v}_{cm} = m_A\vec{v}_A + m_B\vec{v}_B + \cdots.$$ The center of mass of a system moves as if all the mass were located at the center of mass: $$\sum \vec{F}_{ext} = M\vec{a}_{cm}.$$

Conceptual Questions

1: Jumping off a wall

If you fall off a 2-m-high wall, would you prefer to land on concrete or grass? Why?

Solution

SET UP AND SOLVE Your speed at the ground will be the same in both cases. The change in your momentum, or impulse, as you come to rest will also be the same. Given the same impulse, the average force will be less if the time interval is longer. Since you will be in contact with the grass longer as you land, grass is the preferred landing material.

REFLECT This conceptual question illustrates why cushioning is used to reduce the average force exerted in a collision by increasing the duration of the collision. Consider this when you buy your next pair of running shoes or feel your car's padded dashboard.

2: Beanbag versus tennis ball

You wish to close your bedroom door from across the room. You can toss either a beanbag or a tennis ball at the door. (Both have the same mass.) Which should you choose?

Solution

SET UP AND SOLVE Consider how much momentum each object can impart on the door. Both the beanbag and the tennis ball have the same mass, and you give each the same velocity, so their initial momenta are the same. After colliding with the door, the beanbag falls to the floor while the tennis ball bounces back towards you. The beanbag ends with zero momentum while the tennis ball has momentum in the direction opposite its original momentum. The change in momentum for the tennis ball is larger than for the beanbag, so you should use the tennis ball to close the door.

REFLECT Even though both objects have the same initial momentum, we see that the *change* in momentum determines the best choice.

3: Getting off the ice

You're standing on a frictionless ice rink. If you toss your physics book vertically upwards, will you move?

Solution

SET UP AND SOLVE You will not move, since the book carries away no component of momentum parallel to the ice rink. You should toss your physics book horizontally to move along the ice.

REFLECT Is momentum conserved in this case? Initially, there is no momentum. When you toss the book, you and the earth must move in the opposite direction. Since the mass of the earth is extremely large, the velocity imparted to the earth is tiny.

4: Walking in a canoe

You are standing in a canoe. If you walk to the other end of the canoe, what happens to the canoe? Neglect the resistance of the water.

SET UP AND SOLVE No external forces act on you or the canoe (ignoring the resistance of the water). Your center of mass and that of the canoe remain constant. When you walk to the other end of the canoe, the canoe must move in the opposite direction in order to preserve the center of mass.

REFLECT This conceptual question illustrates how the frictional force is necessary for walking. As you walk, the frictional force between your shoes and the canoe pushes you in one direction while pushing the canoe in the opposite direction.

Problems

1: Tossing bubble gum

You throw your bubble gum at a stationary puck on a frictionless air-hockey table. The gum sticks to the puck, and both move away with a velocity of 1.2 m/s. If the puck has mass 0.30 kg and the bubble gum has mass 0.020 kg, find the initial speed of the bubble gum.

Solution

SET UP Figure 8.1 shows the before and after sketches of the situation, including axes. The system consists of the bubble gum and puck. Before the collision, the gum has a velocity v_i and the puck is stationary. After the collision, the gum and puck move away together at velocity v_f. All of the velocities and momenta have only x components. There are no horizontal external forces, so the x component of total momentum is the same before and after the collision.

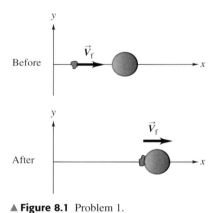

▲ **Figure 8.1** Problem 1.

SOLVE We solve the problem by using conservation of momentum. The initial momentum is that of the gum:

$$P_i = m_g v_i.$$

The final momentum is that of the gum and puck together:

$$P_f = (m_g + m_p)v_f.$$

Momentum is conserved, so we set the momenta equal to each other:

$$m_g v_i = (m_g + m_p)v_f.$$

Solving for v_i yields

$$v_i = \frac{(m_g + m_p)v_f}{m_g} = \frac{((0.020 \text{ kg}) + (0.30 \text{ kg}))(1.2 \text{ m/s})}{(0.020 \text{ kg})} = 19 \text{ m/s}.$$

The initial speed of the bubble gum is 19 m/s.

REFLECT This problem illustrates how we can determine a projectile's velocity by examining its collision with a larger object and applying conservation of momentum.

Was the collision elastic? No, the collision was not elastic: The initial kinetic energy is 3.6 J and the final kinetic energy is 0.23 J. This is an example of a totally inelastic collision, since the masses stuck together after the collision.

2: Rocket motion

A two-stage rocket traveling at 350 m/s through space separates into its two stages, one having twice the mass of the other. If the final velocity of the larger stage is 120 m/s in a direction opposite its initial direction, find the final velocity of the smaller stage.

Solution

SET UP Figure 8.2 shows the before and after sketches of the situation, including axes. The system consists of the two rocket stages. Before the collision, the stages are connected and are moving with velocity v_i to the right. After the collision, stage A (with mass m) moves to the left at velocity $v_{A,f}$ and stage B (with mass $2m$) moves to the right at velocity $-v_{B,f}$ (B must move to the right to conserve momentum.) All of the velocities and momenta have only x components. There are no horizontal external forces, so the x component of total momentum is the same before and after the collision.

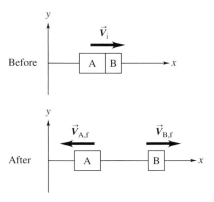

▲ **Figure 8.2** Problem 2.

SOLVE We solve the problem by using conservation of momentum. The initial momentum is that of both stages moving together:

$$P_i = (m_A + m_B)v_i.$$

The final momentum is that of the two stages separated:

$$P_f = m_A v_{A,f} + m_B v_{B,f}.$$

Momentum is conserved, so we set the momenta equal to each other:

$$m_A v_i + m_B v_i = m_A v_{A,f} + m_B v_{B,f}.$$

Solving for $v_{A,f}$ and substituting for the masses gives

$$v_{A,f} = \frac{(m_A + m_B)v_i - m_B v_{B,f}}{m_A} = \frac{(m + 2m)(350 \text{ m/s}) - 2m(-120 \text{ m/s})}{m} = 1300 \text{ m/s}.$$

The final speed of the smaller stage is 1300 m/s.

REFLECT The smaller stage gets a significant increase in velocity after the separation. This results from both its smaller mass and the larger stage moving off in the opposite direction.

3: Colliding pucks in two dimensions

Two pucks collide on a frictionless air-hockey table. Initially, puck *A* is traveling at 3.50 m/s and puck *B* is at rest. After the collision, puck *A* moves away at a speed of 2.50 m/s and an angle of 30.0°. Find the final velocity (both magnitude and direction) of puck *B*. Both pucks have the same mass.

Solution

SET UP Figure 8.3 shows the before and after sketches of the situation, including axes. The system consists of the two pucks. Before the collision, puck *A* moves with velocity v_i to the right. After the collision, puck *A* moves away at velocity $v_{A,f}$, 30.0° above the *x* axis and puck *B* moves away at velocity $v_{B,f}$, at an angle θ below the *x* axis (to conserve momentum). The velocities are not along a single axis, so we will have to solve for both the *x* and *y* components of momentum. There are no horizontal external forces, so the total horizontal momentum is the same before and after the collision.

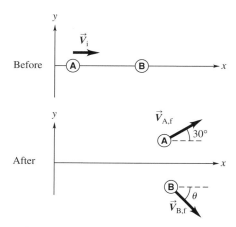

▲ **Figure 8.3** Problem 3.

SOLVE We solve the problem by using conservation of momentum. We start with the *x* components of momentum. We have

$$P_{i,x} = mv_i,$$
$$P_{f,x} = mv_{A,f,x} + mv_{B,f,x} = mv_{A,f}\cos 30.0° + mv_{B,f,x}.$$

The *x* component of momentum is conserved, so we set the expressions for $P_{i,x}$ and $P_{f,x}$ equal to each other:

$$mv_i = mv_{A,f}\cos 30.0° + mv_{B,f,x}.$$

Solving for $v_{B,f,x}$ yields

$$v_{B,f,x} = \frac{mv_i - mv_{A,f}\cos 30.0°}{m} = (3.50 \text{ m/s}) - (2.50 \text{ m/s})\cos 30.0° = 1.34 \text{ m/s}.$$

We follow the same procedure for the *y* components of momentum. The initial *y* component of momentum is zero. The *y* components of final momentum must sum to zero; that is,

$$P_{f,y} = mv_{A,f,y} + mv_{B,f,y} = mv_{A,f}\sin 30.0° + mv_{B,f,x} = 0.$$

Solving for $v_{B,f,x}$ gives

$$v_{B,f,y} = -v_{A,f}\sin 30.0° = -(2.50 \text{ m/s})\sin 30.0° = -1.25 \text{ m/s}.$$

The x and y components of puck B's velocity are 1.34 m/s and -1.25 m/s, respectively. Thus, puck B must travel into the fourth quadrant, as expected. We find the magnitude and direction of B's velocity from

$$v_{B,f} = \sqrt{v_{B,f,x}^2 + v_{B,f,y}^2} = \sqrt{(1.34 \text{ m/s})^2 + (-1.25 \text{ m/s})^2} = 1.83 \text{ m/s},$$

$$\phi = \tan^{-1}\frac{v_{B,f,y}}{v_{B,f,x}} = \tan^{-1}\frac{(-1.25 \text{ m/s})^2}{(1.34 \text{ m/s})} = -43.0°,$$

Puck B's final velocity is 1.83 m/s, directed at an angle of 43.0° below the positive x axis.

REFLECT Solving momentum problems in two dimensions follows from the one-dimensional cases. You just need to remember that momentum is a vector and can have multiple components.

4: Elastic collision in one dimension

Two gliders collide elastically on a frictionless, linear air track. Glider A has mass 0.60 kg and initially moves to the right at 3.0 m/s. Glider B has mass 0.40 kg and initially moves to the left at 4.0 m/s. What are the final velocities of the two gliders after the collision?

Solution

SET UP Figure 8.4 shows the before and after sketches of the situation, including axes. The system consists of the two gliders. Before the collision, glider A moves with velocity $v_{A,i} = 3.0$ m/s to the right and glider B moves with velocity $v_{B,i} = -4.0$ m/s to the left. After the collision, glider A moves away at velocity $v_{A,f}$ and glider B moves away at velocity $v_{B,f}$. All of the velocities and momenta have only x components. There are no horizontal external forces, so the x component of total momentum is the same before and after the collision. This is an elastic collision, so the initial and final kinetic energies are equal. We'll use the relative-velocity relation for elastic collisions in our solution.

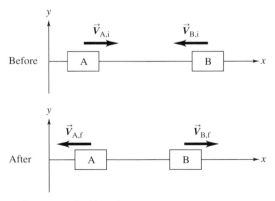

▲ **Figure 8.4** Problem 4.

SOLVE We solve the problem by using conservation of momentum. We have

$$P_i = m_A v_{A,i} + m_B v_{B,i},$$
$$P_f = m_A v_{A,f} + m_B v_{B,f}.$$

Momentum is conserved, so we set the right-hand sides equal to each other:

$$m_A v_{A,i} + m_B v_{B,i} = m_A v_{A,f} + m_B v_{B,f},$$
$$(0.60 \text{ kg})(3.0 \text{ m/s}) + (0.40 \text{ kg})(-4.0 \text{ m/s}) = (0.60 \text{ kg})v_{A,f} + (0.40 \text{ kg})v_{B,f},$$
$$0.60 v_{A,f} + 0.40 v_{B,f} = 0.20 \text{ m/s}.$$

This final equation has two unknowns. We need to apply the relative-velocity relation for elastic collisions to solve it:

$$v_{B,f} - v_{A,f} = -(v_{B,i} - v_{A,i}),$$
$$v_{B,f} - v_{A,f} = -((-4.0 \text{ m/s}) - (3.0 \text{ m/s})) = 7.0 \text{ m/s},$$
$$v_{B,f} = v_{A,f} + 7.0 \text{ m/s}.$$

Substituting the last expression into the equation for conservation of momentum gives

$$0.60 v_{A,f} + 0.40(v_{A,f} + 7.0 \text{ m/s}) = 0.20 \text{ m/s},$$
$$v_{A,f} = (0.20 \text{ m/s}) - (0.40(7.0 \text{ m/s})) = -2.6 \text{ m/s},$$
$$v_{B,f} = v_{A,f} + 7.0 \text{ m/s} = (-2.6 \text{ m/s}) + 7.0 \text{ m/s} = 4.4 \text{ m/s}.$$

After the collision, puck A moves to the left at 2.6 m/s ($v_{A,f}$ is negative) and puck B moves to the right at 4.4 m/s.

REFLECT Both gliders reversed their directions in this elastic collision. Are the kinetic energies equivalent before and after the collision? The initial kinetic energy is

$$K_i = \tfrac{1}{2} m_A v_{A,i}^2 + \tfrac{1}{2} m_B v_{B,i}^2 = \tfrac{1}{2}(0.60 \text{ kg})(3.0 \text{ m/s})^2 + \tfrac{1}{2}(0.40 \text{ kg})(-4.0 \text{ m/s})^2 = 5.9 \text{ J}.$$

The final kinetic energy is

$$K_f = \tfrac{1}{2} m_A v_{A,f}^2 + \tfrac{1}{2} m_B v_{B,f}^2 = \tfrac{1}{2}(0.60 \text{ kg})(-2.6 \text{ m/s})^2 + \tfrac{1}{2}(0.40 \text{ kg})(4.4 \text{ m/s})^2 = 5.9 \text{ J}.$$

Both the initial and final kinetic energies are equivalent, as expected.

Problem Summary

This chapter has augmented our knowledge of energy, forces, and kinematics with our newly formed knowledge of momentum analysis. Momentum analysis shares many of the problem-solving principles we have encountered. In these problems, we

- Identified the general procedure for finding the solution.
- Sketched the situation, including before and after views.
- Identified the momenta, energies, and forces in the system.
- Applied conservation of momentum to the system.
- Applied energy principles, including conservation of energy (when possible).
- Drew free-body diagrams for the objects when appropriate.
- Applied Newton's laws when appropriate.
- Solved the equations through algebra, including substitutions.
- Reflected on the results, thus checking for inconsistencies.

9 ROTATIONAL MOTION

Summary

In this chapter, we will investigate the rotational motion of *rigid bodies,* objects that don't change size or shape as they move. We'll first describe the kinematics of rotation for a rigid body and then examine its rotational kinetic energy. We'll see that these quantities are analogous to linear kinematics and translational kinetic energy. We'll use our new knowledge about rotational motion in the next chapter as we learn how to cause such motion.

Objectives

- Learn to use radians in angular measurements.
- Learn the definition and application of angular displacement, velocity, and acceleration.
- Solve problems involving constant angular acceleration.
- Define moment of inertia and apply it to systems of varying shapes.
- Solve conservation-of-energy problems that include rotational kinetic energy.
- Draw analogies between translational and rotational motion and energy.

Concepts and Equations

Term Description	
Rigid Body	A rigid body is an object that maintains an unchanging size and shape. We neglect squeezing, stretching, and twisting in our analysis of a rigid body.
Radian	In analyzing rotational motion problems, we usually measure angles in radians. An angle θ in radians is the ratio of the arc length s to the radius r: $$\theta = \frac{s}{r}.$$ There are 2π radians in one revolution ($360°$).
Angular Velocity	The average angular velocity describes how a rigid body rotates through an angular displacement $\Delta\theta$ in a time interval Δt: $$\omega_{av} = \frac{\Delta\theta}{\Delta t}.$$ The instantaneous angular velocity is $$\omega = \lim_{\Delta t \to 0} \frac{\Delta\theta}{\Delta t}.$$ The term *angular velocity* refers to the instantaneous angular velocity. All pieces of a rigid object have the same angular velocity at any given instant. We often assign counterclockwise angular displacements and velocities to positive values and clockwise angular displacements and velocities to negative values.
Angular Acceleration	The average angular acceleration is the change in angular velocity $\Delta\omega$ per time interval Δt: $$\alpha_{av} = \frac{\Delta\omega}{\Delta t}.$$ The instantaneous angular acceleration is $$\alpha = \lim_{\Delta t \to 0} \frac{\Delta\omega}{\Delta t}.$$ The term *angular acceleration* refers to the instantaneous angular acceleration. All pieces of a rigid object have the same angular acceleration at any given instant.
Rotation with Constant Angular Acceleration	When an object moves with constant angular acceleration, the angular position, velocity, acceleration, and time are related by $$\theta = \theta_0 + \omega_0 t + \tfrac{1}{2}\alpha t^2,$$ $$\omega = \omega_0 + \alpha t,$$ $$\omega^2 = \omega_0^2 + 2\alpha(\theta - \theta_0),$$ $$\theta - \theta_0 = \tfrac{1}{2}(\omega + \omega_0)t,$$ where θ_0 and ω_0 are the initial values of the angular position and velocity.
Connecting Linear and Angular Quantities	The tangential speed v of a particle rotating in a rigid body at a distance r from the axis of rotation is $$v = \omega r.$$ The particle's acceleration $\vec{a}$ has a tangential component $$a_{tan} = \alpha r$$

and a radial component

$$a_{\text{rad}} = \frac{v^2}{r} = \omega^2 r.$$

Moment of Inertia	The moment of inertia, I, of a body describes how the body's mass is distributed relative to the axis of rotation and is given by $$I = m_A r_A^2 + m_B r_B^2 + m_C r_C^2 + \cdots.$$ Moments of inertia for common shapes are given in Table 9.2 in the textbook.
Energy of Rotating Body	The kinetic energy of a rigid body rotating about a stationary axis is $$K = \tfrac{1}{2} I \omega^2.$$ This quantity is the sum of the kinetic energies of all of the particles that make up the rigid body. The gravitational potential energy of a body in a uniform gravitational field with center of mass at a distance y_{cm} above the reference level $(U = 0)$ is $$U = M g y_{\text{cm}}.$$ When a rigid body undergoes both motion of its center of mass and rotation about an axis through its center of mass, the total kinetic energy of the body is $$K = \tfrac{1}{2} M v_{\text{cm}}^2 + \tfrac{1}{2} I \omega^2.$$

Conceptual Questions

1: Rolling versus sliding down a hill

A ball travels down a hill. Will the ball reach the bottom of the hill faster if it rolls or slides without friction down the hill?

Solution

SET UP AND SOLVE No energy is lost as the ball travels down the hill, so we can use energy conservation to answer this question. At the top of the hill, the ball has gravitational potential energy. As it descends, the gravitational potential energy transforms into kinetic energy. When the ball rolls down the hill, the kinetic energy is shared between translational kinetic energy and rotational kinetic energy. When the ball slides down the hill without friction, the gravitational potential energy transforms into translational kinetic energy. There is no rotational kinetic energy in this case. Therefore, more energy is transformed into translational kinetic energy if the ball slides without friction than if it rolls down the hill. Since the ball acquires more translational kinetic energy, its velocity is higher and it reaches the bottom faster when it slides without friction.

REFLECT This question shows how we must include both translational and rotational kinetic energy in our analyses of energy. We'll practice using rotational kinetic energy in the problem section.

2: Comparing moments of inertia

A light rod of length L has two lead weights of mass M attached to it, one at each end. How does the system's moment of inertia compare when the rod is spun about an axis at its center as opposed to when it is spun around a point one-quarter along its length?

Solution

SET UP AND SOLVE The moment of inertia of an object is

$$I = m_A r_A^2 + m_B r_B^2 + m_C r_C^2 + \cdots.$$

The moment of inertia when the axis is at the center of the rod is then

$$I_{\text{center axis}} = M\left(\frac{L}{2}\right)^2 + M\left(\frac{L}{2}\right)^2 = M\left(\frac{L^2}{4}\right) = \tfrac{1}{4}ML^2,$$

since each mass is located halfway along the rod's length from the axis. When the axis is one-quarter along the length of the rod, the moment of inertia is

$$I_{\text{1/4 along rod}} = M\left(\frac{L}{4}\right)^2 + M\left(\frac{3L}{4}\right)^2 = M\left(\frac{L^2}{16}\right) + M\left(\frac{9L^2}{16}\right) = \tfrac{10}{16}ML^2,$$

since one mass is $\frac{1}{4}L$ from the axis and the other is $\frac{3}{4}L$ from the axis. The moment of inertia when the rod is spun one-quarter along its length is five times larger than the moment of inertia when the rod is spun at the center.

REFLECT The moment of inertia depends on both the mass and the location of the mass. Since it is the *square* of the distance from the axis, and not the distance itself, that enters into the equation, moving the axis has profound changes in moment of inertia.

Practice Question: What axis of rotation provides the largest moment of inertia? *Answer:* The axis located at one end of the rod gives the largest moment of inertia, $I = ML^2$.

3: Rolling up a ramp

A solid sphere and a thin-walled sphere roll without slipping along a horizontal surface. The two spheres roll with the same translational speed. The surface leads to a ramp. Which sphere rises to the greatest height before momentarily stopping?

Solution

SET UP AND SOLVE We'll use conservation of energy and ignore air drag. Both spheres have initial translational and kinetic energies that transform completely into gravitational potential energy when they stop momentarily on the ramp. The sphere with the greatest initial total kinetic energy will rise to the greatest height. The initial kinetic energy for each sphere is

$$K_i = \tfrac{1}{2}m_{\text{sphere}}v_{\text{cm}}^2 + \tfrac{1}{2}I_{\text{sphere}}\omega^2.$$

Since both spheres roll without slipping, their angular velocity is related to their center of mass by

$$\omega = \frac{v_{\text{cm}}}{R},$$

where R is the radius of the sphere. The moment of inertia of the solid sphere is $I_{\text{solid}} = 2/5\,m_{\text{solid}}R_{\text{solid}}^2$. The kinetic energy of the solid sphere is then

$$K_{\text{solid}} = \tfrac{1}{2}m_{\text{solid}}v_{\text{cm}}^2 + \tfrac{1}{2}I_{\text{solid}}\omega^2 = \tfrac{1}{2}m_{\text{solid}}v_{\text{cm}}^2 + \tfrac{1}{2}\tfrac{2}{5}m_{\text{solid}}R_{\text{solid}}^2\left(\frac{v_{\text{cm}}}{R_{\text{solid}}}\right)^2 = \tfrac{7}{10}m_{\text{solid}}v_{\text{cm}}^2.$$

The moment of inertia of the thin-walled sphere (shell) is $I_{\text{shell}} = 2/3\,m_{\text{shell}}R_{\text{shell}}^2$. The kinetic energy of the shell is then

$$K_{\text{shell}} = \tfrac{1}{2}m_{\text{shell}}v_{\text{cm}}^2 + \tfrac{1}{2}I_{\text{shell}}\omega^2 = \tfrac{1}{2}m_{\text{shell}}v_{\text{cm}}^2 + \tfrac{1}{2}\tfrac{2}{3}m_{\text{shell}}R_{\text{shell}}^2\left(\frac{v_{\text{cm}}}{R_{\text{shell}}}\right)^2 = \tfrac{5}{6}m_{\text{shell}}v_{\text{cm}}^2.$$

Since the final gravitational potential energy depends on mass $(U = mgh)$, the masses will cancel and we can compare the leading fractions in the kinetic-energy terms to determine which sphere has the greatest initial kinetic energy. We see that the shell has more initial kinetic energy; therefore, the thin-walled sphere rises to the greatest height.

REFLECT It is interesting to find that the results don't depend on either the mass or radius of the two spheres. The results depend only on how the mass is distributed in the object.

Problems

1: Constant angular acceleration in a pottery wheel

A pottery wheel is rotating with an initial angular velocity ω_0 when the wheel's drive motor is turned on. The wheel increases to a final angular velocity of 125 rpm while making 30.0 revolutions in 25.0 seconds. Find the initial angular velocity and the angular acceleration of the wheel, assuming that the latter is constant. The pottery wheel is essentially a cylinder rotating about a vertical axis driven by a motor.

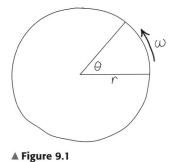

▲ **Figure 9.1**

Solution

SET UP Figure 9.1 shows a sketch of the pottery wheel. Since the wheel exhibits constant angular acceleration, we can use the constant angular acceleration relations. We'll use radians and seconds as our units for consistency and convert the given quantities.

SOLVE We are given the final angular velocity, the angular displacement, and the time. We need to find the initial angular velocity and angular acceleration. None of the constant angular acceleration expressions allow us to solve for both unknowns at once, so we'll solve for the initial angular velocity first and then use the results to solve for the angular acceleration.

The angular displacement can be written in terms of the average angular velocity and the time interval as

$$\theta - \theta_0 = \tfrac{1}{2}(\omega + \omega_0)t.$$

Rearranging terms to find the initial angular velocity gives

$$\omega_0 = \frac{2(\theta - \theta_0)}{t} - \omega.$$

The quantity $(\theta - \theta_0)$ is our angular displacement, 30.0 revolutions. The quantity ω is the final angular velocity, 125 rpm. We convert the revolutions to radians by multiplying by $(2\pi \text{ rad/rev})$ and the rpm to radians/second by multiplying by $(2\pi \text{ rad/rev})(1 \text{ min/60 s})$. Solving for ω_0 yields

$$\omega_0 = \frac{2(30.0 \text{ rev})(2\pi \text{ rad/rev})}{25.0 \text{ s}} - (125 \text{ rpm})(2\pi \text{ rad/rev})(1 \text{ min/60 s}) = 1.99 \text{ rad/s}.$$

The initial angular velocity is 1.99 rad/s, or 19.0 rpm. To find the angular acceleration, we use the relationship between angular velocity, angular acceleration, and time:

$$\omega = \omega_0 + \alpha t.$$

Solving for the angular acceleration, we obtain

$$\alpha = \frac{\omega - \omega_0}{t} = \frac{(125 \text{ rpm})(2\pi \text{ rad/rev})(1 \text{ min/60 s}) - (1.99 \text{ rad/s})}{25.0 \text{ s}} = 0.444 \text{ rad/s}^2.$$

The angular acceleration is 0.444 rad/s².

REFLECT This problem reminds us of our constant linear acceleration problems we first encountered in Chapter 2. The same problem-solving strategy applies to this problem: Draw a diagram, check for constant acceleration, find one or more equations that can be used to solve for the unknowns, and reflect upon the results. As we've seen here, we may need to use more than one equation for the solution, and we must watch our units carefully.

2: Energy in a wheel–stone system

A thin light string is wrapped around the rim of a spoked wheel that can rotate without friction around its center axle. An 8.00 kg stone is attached to the end of the string as shown in Figure 9.2. If the stone is released from rest, how far does it travel before attaining a speed of 4.80 m/s? The spoked wheel is made of a central hub (a solid uniform cylinder of radius 7.50 cm and mass 22.0 kg) attached to a rim (a thin-walled hollow cylinder of radius 30.0 cm and mass 12.0 kg) by spokes of negligible mass.

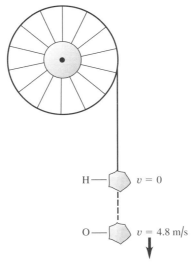

▲ **Figure 9.2**

Solution

SET UP There are no nonconservative forces in this problem, so we'll use conservation of energy. Initially, there is only gravitational potential energy. When the stone is released, its gravitational potential energy transforms into kinetic energy of the stone and wheel. We'll set the origin at the point where the stone reaches a speed of 4.80 m/s; the starting position is a distance H above the origin, as we see in Figure 9.2. We'll find the moment of inertia by combining the moment of inertia of a solid cylinder with that of a thin-walled hollow cylinder.

SOLVE Energy conservation relates the initial and final energies:

$$K_i + U_i = K_f + U_f.$$

Initially, there is only U_{grav}. At the origin, the gravitational potential energy has transformed into the kinetic energies of the stone and wheel:

$$U_{grav} = K_{stone} + K_{wheel}.$$

Replacing the energies yields

$$m_{stone}gH = \tfrac{1}{2}m_{stone}v^2 + \tfrac{1}{2}I_{wheel}\omega^2.$$

   We need to find the moment of inertia of the wheel and its angular velocity when the stone reaches its final velocity. The moment of inertia of the wheel is the algebraic sum of the moment of inertia of the central hub and the moment of inertia of the outer-rim cylinder. Using Table 9.2 of the text, we find the total moment of inertia of the wheel:

$$I_{wheel} = I_{solid\ cylinder} + I_{thin\text{-}walled\ cylinder} = \tfrac{1}{2}M_{hub}R_{hub}^2 + M_{rim}R_{rim}^2.$$

The speed of the stone is the tangential speed of the wheel, so we can find the angular speed of the wheel from

$$\omega = \frac{v}{R_{rim}}.$$

Substituting these two expressions into the energy relation

$$m_{stone}gH = \tfrac{1}{2}m_{stone}v^2 + \tfrac{1}{2}\left(\tfrac{1}{2}M_{hub}R_{hub}^2 + M_{rim}R_{rim}^2\right)\left(\frac{v}{R_{rim}}\right)^2.$$

Solving for H

$$H = \frac{1}{m_{stone}g}\left[\tfrac{1}{2}m_{stone}v^2 + \tfrac{1}{2}\left(\tfrac{1}{2}M_{hub}R_{hub}^2 + M_{rim}R_{rim}^2\right)\left(\frac{v}{R_{rim}}\right)^2\right],$$

or

$$H = \frac{1}{(8.00\ \text{kg})(9.8\ \text{m/s}^2)} \cdot$$
$$\left[\tfrac{1}{2}(8.00\ \text{kg})(4.80\ \text{m/s})^2 + \tfrac{1}{2}\left(\tfrac{1}{2}(22.0\ \text{kg})(0.0750\ \text{m})^2 + (12.0\ \text{kg})(0.300\ \text{m})^2\right)\left(\frac{4.80\ \text{m/s}}{0.300\ \text{m}}\right)^2\right]$$
$$= 3.04\ \text{m}.$$

When the stone falls 3.04 m, its speed will be 4.80 m/s.

REFLECT If the stone had fallen freely, it would have attained its final speed when $h = v/\sqrt{2g}$ (1.08 m). Why does it take almost three times the distance to reach the final speed? Looking at the energy conservation equation, we see that the energy is shared between the kinetic energies of the stone and wheel. Roughly two-thirds of the energy goes into the kinetic energy of the wheel.

Practice Problem: Repeat the preceding problem with only the inner hub of the wheel. *Answer:* 1.28 m.

3: Energy in a falling cylinder

A thin, light string is wrapped around a solid, uniform cylinder of mass M and radius R as shown in Figure 9.3. The string is held stationary and the cylinder is released from rest. What is the cylinder's radius if the cylinder reaches an angular speed of 350.0 rpm after it falls 3.00 m?

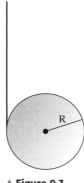

▲ **Figure 9.3**

Solution

SET UP Gravity and tension are the only forces acting in this problem, so energy is conserved. Initially, there is only gravitational potential energy. When the cylinder is released, the gravitational potential energy transforms into rotational and translational kinetic energy. We'll set the origin 3.00 m below the initial position.

SOLVE Energy conservation relates the initial and final energies:

$$K_i + U_i = K_f + U_f.$$

Initially, there is only U_{grav}. At the origin, the gravitational potential energy has transformed into the total kinetic energy of the cylinder:

$$U_{\text{grav}} = K_{\text{translational}} + K_{\text{rotational}}.$$

Replacing the energies, we obtain

$$Mgh = \tfrac{1}{2}Mv^2 + \tfrac{1}{2}I\omega^2.$$

Now, recall that the moment of inertia for a solid uniform cylinder is $\tfrac{1}{2}MR^2$. (See Table 9.2 in the text.) The speed of the cylinder is the tangential speed of the wheel, so we can find the angular speed of the wheel from

$$v = \omega R.$$

Substituting these expressions into the energy relation gives

$$Mgh = \tfrac{1}{2}Mv^2 + \tfrac{1}{2}I\omega^2 = \tfrac{1}{2}M(\omega R)^2 + \tfrac{1}{2}\left(\tfrac{1}{2}MR^2\right)\omega^2 = \tfrac{3}{4}M(\omega R)^2.$$

Solving for R yields

$$R = \frac{\sqrt{\tfrac{4}{3}gh}}{\omega} = \frac{\sqrt{\tfrac{4}{3}(9.80 \text{ m/s}^2)(3.00 \text{ m})}}{(350 \text{ rpm})(2\pi \text{ rad/rev})(1 \text{ min/60 s})} = 0.171 \text{ m}.$$

The cylinder's radius is 17.1 cm.

REFLECT We see that the mass of the cylinder cancels in the conservation-of-energy equation. The results apply to a cylinder of any mass.

Practice Problem: Repeat the preceding problem with a thin hoop replacing the cylinder. *Answer:* 14.8 cm.

4: Cylinder rolling down a ramp

A uniform hollow cylinder of mass 6.5 kg, inner radius 0.13 m, and outer radius 0.25 m approaches a flat ramp and rolls up the ramp without slipping. The ramp is inclined at 20.0°. How far along the ramp does the cylinder roll before stopping if its initial forward speed is 12.0 m/s?

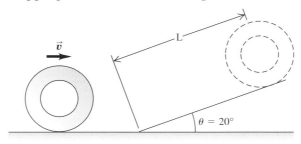

▲ **Figure 9.4**

Solution

SET UP Figure 9.4 shows a sketch of the problem. There are no nonconservative forces (ignoring air drag), so we use energy conservation. Initially, there is translational and rotational kinetic energy. When the cylinder stops momentarily at the top of the ramp, the kinetic energy has totally transformed to gravitational potential energy. We'll set the origin at the base of the ramp.

SOLVE Energy conservation relates the initial and final energies:

$$K_i + U_i = K_f + U_f.$$

Initially, the cylinder has kinetic energy. At its highest point, the cylinder has only gravitational potential energy:

$$K_{\text{translational}} + K_{\text{rotational}} + 0 = 0 + U_{\text{grav}}.$$

Replacing the energies yields

$$\tfrac{1}{2}Mv^2 + \tfrac{1}{2}I\omega^2 = Mgy.$$

Recall that the moment of inertia of a hollow uniform cylinder is $1/2M(R_1^2 + R_2^2)$. (See Table 9.2 in the text.) The speed of the cylinder is the tangential speed of the wheel at the outer radius, so we can replace the speed with

$$\omega = \frac{v}{R_2}.$$

Substituting these expressions into the energy relation gives

$$\tfrac{1}{2}Mv^2 + \tfrac{1}{2}I\omega^2 = \tfrac{1}{2}Mv^2 + \tfrac{1}{2}\left(\tfrac{1}{2}M(R_1^2 + R_2^2)\right)\left(\frac{v}{R_2}\right)^2 = Mgy.$$

Solving for y, the maximum vertical height, we obtain

$$y = \frac{v^2}{g}\left[\frac{1}{2} + \frac{1}{4}\left(\frac{R_1^2 + R_2^2}{R_2^2}\right)\right] = \frac{(12.0\ \text{m/s})^2}{(9.8\ \text{m/s}^2)}\left[\frac{1}{2} + \frac{1}{4}\left(\frac{(0.13\ \text{m})^2 + (0.25\ \text{m})^2}{(0.25\ \text{m})^2}\right)\right] = 12.0\ \text{m}.$$

The cylinder's maximum vertical height is 12.0 m. To find the distance along the ramp, we use the sine relation:

$$L = \frac{y}{\sin 20°} = \frac{(12.0\ \text{m})}{\sin 20°} = 35.1\ \text{m}.$$

The hollow cylinder rolls 35.1 m along the ramp.

REFLECT How far up the ramp would the cylinder travel without friction? It would move 21.5 m along the ramp without friction. This problem shows how the added initial rotational kinetic energy results in a greater distance traveled along the ramp.

DYNAMICS OF ROTATIONAL MOTION

Summary

In this chapter, we will investigate the dynamics of rotational motion to learn what gives an object angular acceleration. We will define torque—the turning or twisting effort of a force—and learn how to apply it to both equilibrium and nonequilibrium situations. Work and power for rotating systems will also be investigated. Angular momentum will be introduced and become the basis of an important, new conservation law. Finally, we will recast our rotational dynamics in vector form. Our linear dynamics foundation developed throughout the text will help build intuition about rotational dynamics.

Objectives

- Learn the definitions of torque and angular momentum.
- Identify torques acting on a body.
- Learn the equation of motion for rotational systems and apply it to problems.
- Apply conservation of angular momentum to problems.
- Apply work and power to rotational dynamics problems.
- Combine linear and rotational equilibrium conditions and apply those conditions to problems.
- Reevaluate angular momentum in vector quantities.

Concepts and Equations

Term	Description
Torque	Torque is the measure of the tendency of a force to cause or change rotational motion about a chosen axis. The magnitude of torque is the magnitude of the force (F) times the moment arm (l)—the perpendicular distance between the axis and line of force: $$\tau = Fl.$$ For a force $\vec{F}$ applied at point P, and for vector $\vec{r}$ from the chosen axis to point P, the torque is given by $$\tau = F_{\tan}r = rF\sin\phi,$$ where ϕ is the angle between $\vec{F}$ and $\vec{r}$. The SI unit of torque is the newton-meter (Nm). Our convention is that positive torques are counterclockwise and negative torques are clockwise.
Torque and Angular Acceleration	The net torque on a rigid body is proportional to the body's angular acceleration and its moment of inertia: $$\sum\tau = I\alpha.$$
Work and Power with a Constant Torque	The work done by a constant torque is the product of the torque and the angular displacement of the body: $$W = \tau\Delta\theta.$$ The power provided by a constant torque is the product of the torque and the angular velocity of the body: $$P = \tau\omega.$$
Angular Momentum	The angular momentum L of a rigid body with respect to an axis is the product of the moment of inertia I about the axis and the angular velocity ω: $$L = I\omega.$$ Our convention is that counterclockwise rotations have positive L and clockwise rotations have negative L.
Torque and Angular Momentum	The rate of change of an object's angular momentum with respect to any axis is the sum of the torques acting on the object about the axis: $$\sum\tau = \frac{\Delta L}{\Delta t}.$$
Conservation of Angular Momentum	If the sum of all torques acting on a system is zero, the total angular momentum of the system is conserved and remains constant.
Equilibrium of a Rigid Body	No net force and no net torque acts on a rigid body in equilibrium: $$\sum\vec{F} = 0 \text{ and } \sum\tau = 0.$$
Vector Nature of Angular Quantities	Torque, angular velocity, and angular momentum are vector quantities whose magnitudes are their scalar values and whose directions are given by the right-hand rule. Torque, angular velocity, and angular momentum are related by $$\vec{L} = I\vec{\omega} \text{ and } \sum\vec{\tau} = \frac{\Delta\vec{L}}{\Delta t}.$$

Conceptual Questions

1: Ranking torques

In the following diagrams, each rod pivots about the indicated axis with the indicated force:

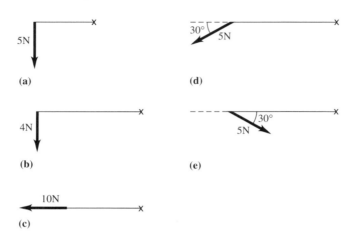

Rank the diagrams in order of increasing torque.

Solution

SET UP AND SOLVE Torque is given by $\tau = rF\sin\phi$, so we must examine the magnitude of the force, where the force is applied, and the direction in which the force is applied. Comparing (a) and (b), we see that both forces act in the same direction, the force is larger in (a), and the moment arm is larger in (b). We estimate the moment arm in (a) as half the moment arm in (b). Since the force in (a) is less than double the force in (b), the torque in (b) is greater than that in (a). No torque is generated in (c), since the force acts along the moment arm. The force and moment arm are the same in (d) and (e), but the directions are different. However, since both forces act 30° from the horizontal, the components of the forces perpendicular to the moment arm are the same for both, and the torque in (d) and (e) are the same. The vertical component of the force in (d) and (e) is 2.5 N, which is less than the force in (b) and less than half of the force in (a). Diagrams (d) and (e) both have less torque than diagrams (a) and (b).

A ranking of the diagrams in order of increasing torque is therefore (c), $(d) = (e)$, (a), (b).

REFLECT Torque is the most complicated quantity that we have experienced thus far. It depends on the magnitude and direction of the force responsible for it, as well as where that force acts relative to the rotation axis. Gaining intuition about torque will help guide you through problems.

2: Massless pulleys versus pulleys with mass

Why were massless pulleys used in problems from the previous chapters?

Solution

SET UP AND SOLVE Consider the net torque acting on a pulley with a rope resting on the pulley. The left segment of rope has a tension T_L and the right segment of rope has a tension T_R. The net torque is then

$$\sum \tau = T_R R - T_L R = I\alpha.$$

Each tension creates a torque TR, since the rope will be perpendicular to the radius. Massless pulleys have no moment of inertia (since their mass is zero); therefore, the net torque acting on the pulley is zero. If the net torque is zero, the torques in each segment must be equal and the tensions are equal in each rope.

When we include the pulley's mass, the pulley has a moment of inertia. If the pulley accelerates, there must be a difference in the left and right torques, and the tensions must also be different. Only when the angular acceleration is zero are the two tensions equal.

Massless pulleys were used in earlier chapters to avoid having to include torque in our problems. Using massless pulleys simplified our analysis and let us focus on learning about forces.

REFLECT From now on, we must assume that the tensions in segments may vary and therefore we must identify each segment's tension separately.

How does the acceleration of each segment of rope compare when we include a pulley's mass? There is no change: Objects connected by the rope are constrained to have the same magnitude of acceleration.

3: Spinning on a roundabout

You are standing at the center of a rotating playground roundabout (a round, horizontal plate that spins about its center axis). As you move to the edge, will your angular speed increase, decrease, or stay the same?

Solution

SET UP AND SOLVE We ignore any friction in the roundabout to simplify our analysis. Without friction, there is no torque to slow the roundabout, so angular momentum is conserved. As you move to the edge, the moment of inertia of the system increases. (Your mass moves to a greater radius.) For angular momentum to be conserved, the angular speed must be reduced.

REFLECT Like linear momentum and energy, angular momentum is a conserved quantity. We could solve this problem numerically by picking initial and final angular momenta before and after you moved and setting them equal to each other.

Problems
1: Balancing a food tray

A waiter balances a tray of food on his hand. On the tray is a 0.40 kg drink and a 2.0 kg lobster dinner. The drink is placed 6.5 cm from one edge of the tray, and the lobster dinner is placed 8.0 cm from the opposite edge. The tray has a mass of 1.2 kg and a diameter of 42 cm. Where should the waiter hold the tray so that it doesn't tip over?

Solution

SET UP Figure 10.1 shows a sketch and a free-body diagram of the food tray. The forces on the tray include the force of the waiter's hand holding the tray, the weight of the tray, and the normal forces due to the lobster dinner and drink. To find the location of the hand, we need to consider the torque acting on the tray. The tray is to be in equilibrium to prevent falling, so the net torque must be zero around the chosen axis. We'll take the axis to be the left edge, marked by an X.

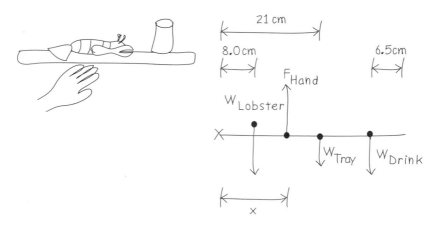

▲ **Figure 10.1**

SOLVE When we choose the left edge as the axis, we note that four torques act on the tray, corresponding to the four forces on the tray. Each of the four forces is applied perpendicular to the moment arm (the plane of the tray); each torque is the magnitude of the force times the distance from the axis. We'll use the standard convention of counterclockwise torques being positive. Since the tray is in equilibrium, the net torque on it is zero:

$$\sum \tau = \tau_{\text{tray}} + \tau_{\text{lobster}} + \tau_{\text{drink}} + \tau_{\text{hand}} = 0.$$

Writing the four torques explicitly, we have

$$\sum \tau = \tau_{\text{tray}} + \tau_{\text{lobster}} + \tau_{\text{drink}} + \tau_{\text{hand}} = -m_{\text{tray}}g x_{\text{tray}} - n_{\text{lobster}} x_{\text{lobster}} - n_{\text{drink}} x_{\text{drink}} + n_{\text{hand}} x_{\text{hand}} = 0$$

The first term is the torque due to the weight of the tray; the moment arm is half the diameter of the tray (at its center of mass). The second and third terms are the torques due to the normal forces of the lobster dinner and drink, respectively. The normal force is equal to the weight of the objects, and the moment arm is the distance from the left edge of the tray. The first three terms are negative, since they are all in the clockwise direction. The last term is the torque due to the normal force of the waiter's hand and is the only positive term, because it is counterclockwise. We need to find this normal force to solve the problem. We find it by using Newton's first law:

$$\sum F = 0.$$

In the vertical direction, four forces act on the tray. We have

$$\sum F = -m_{\text{tray}}g - n_{\text{lobster}} - n_{\text{drink}} + n_{\text{hand}} = 0.$$

Solving yields

$$n_{\text{hand}} = (m_{\text{tray}} + m_{\text{lobster}} + m_{\text{drink}})g = ((1.2 \text{ kg}) + (2.0 \text{ kg}) + (0.40 \text{ kg}))(9.8 \text{ m/s}^2) = 35.3 \text{ N}.$$

We can now solve for the location of the waiter's hand:

$$x_{\text{hand}} = \frac{m_{\text{tray}}g x_{\text{tray}} + m_{\text{lobster}}g x_{\text{lobster}} + m_{\text{drink}}g x_{\text{drink}}}{n_{\text{hand}}}$$

$$= \frac{((1.2 \text{ kg})(0.21 \text{ m}) + (2.0 \text{ kg})(0.080 \text{ m}) + (0.40 \text{ kg})(0.42 \text{ m} - 0.065 \text{ m}))(9.8 \text{ m/s}^2)}{(35.3 \text{ N})}$$

$$= 0.15 \text{ m}.$$

The waiter should hold the tray 15 cm from the left edge (closest to the lobster dinner).

REFLECT This problem required us to apply both the equilibrium torque and equilibrium force conditions. You should be quite familiar with the equilibrium force condition and should only need to gain expertise in understanding torque to solve similar problems. Experience will show us that we may be able to simplify similar problems by picking an axis that coincides with where a force is applied. This choice will reduce the number of torques in the problem. In the current problem, for example, we could have chosen the axis to be at the location of the lobster dinner, thus removing the torque due to the normal force of the lobster dinner.

2: Tension in string attached to a falling cylinder

A thin, light string is wrapped around the outer rim of a uniform, hollow cylinder of mass 12.0 kg, inner radius 15.0 cm, and outer radius 30.0 cm. The cylinder is released from rest. What is the tension in the string as the cylinder falls?

Solution

SET UP Figure 10.2 shows a sketch and a free-body diagram of the situation. As the cylinder falls, it will accelerate downwards and rotate about its central axis. In falling, the cylinder will rotate faster and undergo angular acceleration. The cylinder has both a net force and a net torque acting on it, and we will need to apply Newton's second law and its rotational analog to solve for the tension.

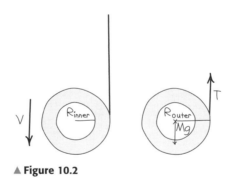

▲ **Figure 10.2**

SOLVE We first apply Newton's second law to the translational motion of the center of mass in the vertical direction. The only forces acting in the vertical direction are gravity and tension. We have

$$\sum F_y = Mg - T = Ma_{cm,y}.$$

The moment of inertia of the hollow cylinder is $I_{cm} = \frac{1}{2}M(R_{inner}^2 + R_{outer}^2)$. The one torque acting on the cylinder as it rotates about its central axis is due to the tension force. (Gravity acts on the center of mass of the cylinder and creates no torque about the central axis.) The torque acts perpendicular to the outer radius. Thus,

$$\sum \tau = TR_{outer} = I_{cm}\alpha = \frac{1}{2}M(R_{inner}^2 + R_{outer}^2)\alpha.$$

We can equate the two accelerations, since the cylinder falls without slipping:

$$a_{cm,y} = \alpha R_{outer}.$$

Solving for the acceleration of the center of mass in the first equation yields

$$a_{cm,y} = \frac{Mg - T}{M}.$$

Substituting the last two equations into the result for the torque gives

$$TR_{\text{outer}} = \tfrac{1}{2}M(R_{\text{inner}}^2 + R_{\text{outer}}^2)\frac{a_{\text{cm},y}}{R_{\text{outer}}} = \tfrac{1}{2}M(R_{\text{inner}}^2 + R_{\text{outer}}^2)\frac{Mg - T}{MR_{\text{outer}}},$$

$$T = \frac{(R_{\text{inner}}^2 + R_{\text{outer}}^2)Mg}{R_{\text{inner}}^2 + 3R_{\text{outer}}^2} = \frac{((15.0\text{ cm})^2 + (30.0\text{ cm})^2)(12.0\text{ kg})(9.8\text{ m/s}^2)}{(15.0\text{ cm})^2 + 3(30.0\text{ cm})^2} = 45.2\text{ N}.$$

The tension in the string is 45.2 N.

REFLECT This problem resembles our earlier Newton's law problems, but with the addition of torque. Once torque is included, the problem becomes a relatively straightforward algebraic one. We also see that the tension in the string is less than that for a stationary cylinder. If the cylinder were stationary, the tension would have been 118 N.

Practice Problem: At what rate does the cylinder accelerate? *Answer:* 6.03 m/s^2.

3: Acceleration and tension of two blocks connected by a pulley.

Two blocks are connected to each other by a light cord passing over a pulley as shown in Figure 10.3. Block A has mass 5.00 kg and block B has mass 4.00 kg. The pulley has mass 8.00 kg and radius 4.00 cm. Find the acceleration of the blocks and the tensions in the horizontal and vertical segments of the cord. Assume that the pulley is a solid, uniform disk and there is no friction between block A and the table.

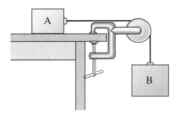

▲ **Figure 10.3**

Solution

SET UP Figure 10.4 shows the free-body diagram for the two blocks and the pulley. The forces on the blocks include tension, gravity, and the normal force (block A). We assume that the tensions of the two segments are not equal and label them T_A and T_B. The two tension forces lead to two torques acting on the pulley. (The axis of rotation is the center of the pulley.) We'll apply the net-force and net-torque equations to solve the problem. The accelerations of both blocks are the same, as we saw in Chapter 5.

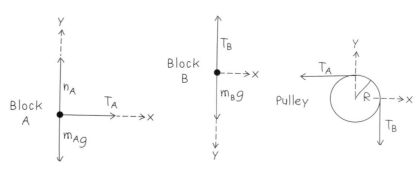

▲ **Figure 10.4**

SOLVE We first apply Newton's second law to each block. Block A (with mass m_A) accelerates in the x direction due to tension T_A, so

$$\sum F_x = T_A = m_A a.$$

Gravity and tension T_B act on block B (with mass m_B), and that block accelerates at the same rate as block A in the y direction, so

$$\sum F_y = m_B g + (-T_B) = m_B a.$$

As block B falls, the pulley's rotational speed increases. The net torque on the pulley is

$$\sum \tau = \tau_A - \tau_B = I\alpha,$$

where we have taken the counterclockwise torque as positive and clockwise torque as negative. The moment of inertia of a uniform cylinder is $I = \frac{1}{2}MR^2$. We assume that the cord doesn't slip on the pulley, so we relate the angular acceleration of the pulley to the tangential acceleration of the cord (a):

$$\alpha = -\frac{a}{R}.$$

We included a minus sign in the equation since the pulley rotates clockwise (negative according to our convention). The tension force acts perpendicular to the moment arm, so the torques are simply TR. Rewriting the net torque, we have

$$\sum \tau = \tau_A - \tau_B = T_A R - T_B R = I\alpha = \frac{1}{2}MR^2\left(-\frac{a}{R}\right).$$

Simpifying yields

$$T_B - T_A = \frac{1}{2}Ma.$$

Our second-law equations are used to replace the tensions:

$$m_B g - m_B a - m_A a = \frac{1}{2}Ma.$$

Solving for the acceleration gives

$$a = \frac{m_B g}{m_B + m_A + \frac{1}{2}M} = \frac{(4.00 \text{ kg})(9.8 \text{ m/s}^2)}{(4.00 \text{ kg}) + (5.00 \text{ kg}) + \frac{1}{2}(8.00 \text{ kg})} = 3.02 \text{ m/s}^2.$$

Using this result to find the two tensions, we get

$$T_A = m_A a = (5.00 \text{ kg})(3.02 \text{ m/s}^2) = 15.1 \text{ N},$$
$$T_B = m_B g - m_B a = (4.00 \text{ kg})(9.80 \text{ m/s}^2) - (4.00 \text{ kg})(3.02 \text{ m/s}^2) = 27.1 \text{ N}.$$

The blocks accelerate at 3.02 m/s^2, the tension in the horizontal segment of the cord is 15.1 N, and the tension in the vertical segment of the cord is 27.1 N.

REFLECT The tensions in the two segments of the cord differ by almost a factor of two. Recall from our problems in Chapter 5 that the tension was constant in both segments of the cord. What causes this difference? The answer is that this problem includes the pulley's mass, resulting in some energy spent on increasing the pulley's angular velocity, leaving less energy available for the blocks. This situation also illustrates why the first pulleys we encountered were massless.

4: Solid cylinder rolling down ramp

A solid cylinder rolls down an incline of 40° without slipping. Find the acceleration and minimum coefficient of friction needed to prevent slipping.

Solution

SET UP Figure 10.5 shows a sketch and free-body diagram for the cylinder on the incline. Gravity, the normal force, and the frictional force act on the cylinder. If we set the axis of rotation at the center of the cylinder, a torque due to friction acts on the cylinder. We'll apply the net-force and net-torque equations to solve the problem.

A rotated coordinate system is included in the free-body diagram to simplify our analysis.

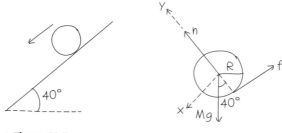

▲ **Figure 10.5**

SOLVE We first apply Newton's second law to the translational motion along the x axis:

$$\sum F_x = Mg\sin\theta - f_s = Ma_{cm,x}.$$

The equation of motion for rotation about that axis is

$$\sum \tau = f_s R = I_{cm}\alpha = \tfrac{1}{2}MR^2\alpha,$$

where we included the moment of inertia $\left(I_{cm} = \tfrac{1}{2}MR^2\right)$. The translational and rotational accelerations are related by $a_{cm} = \alpha R$, since the cylinder rolls without slipping. Combining and writing the second equation in terms of f_s yields

$$f_s = \tfrac{1}{2}Ma_{cm,x}.$$

We use this result in the first equation and solve for the acceleration:

$$Mg\sin\theta - \tfrac{1}{2}Ma_{cm,x} = Ma_{cm,x},$$
$$a_{cm,x} = \tfrac{2}{3}g\sin\theta = \tfrac{2}{3}(9.80 \text{ m/s}^2)\sin 40° = 4.20 \text{ m/s}^2.$$

To find the minimum coefficient of static friction, we use the equilibrium equation along the y axis:

$$\sum F_y = n - Mg\cos\theta = 0.$$

The friction force is then

$$f_s = \mu_s n = \mu_s Mg\cos\theta = \tfrac{1}{2}Ma_{cm,x}.$$

Solving for μ_s gives

$$\mu_s = \frac{\frac{1}{2}a_{cm,x}}{g\cos\theta} = \frac{\frac{1}{2}(4.20 \text{ m/s}^2)}{(9.8 \text{ m/s}^2)\cos 40°} = 0.280.$$

The cylinder accelerates at 4.20 m/s^2, and the minimum coefficient of static friction to prevent slipping is 0.280.

REFLECT This problem is a straightforward application of net-force and net-torque problem-solving techniques. We see that neither the mass nor the radius of the cylinder affects the results, which are valid for *any* cylinder rolling down a 40° incline.

Problem Summary

These first 10 chapters of the book form the basis of kinematic and dynamic problem-solving techniques. Future chapters expand on this basis to include additional forces and forms of energy, but still utilize the same problem-solving techniques. Our problem-solving method continues to consistently include

- Identifying the general procedure to find the solution.
- Sketching the situation when no figure is provided.
- Identifying the forces and torques acting on the bodies.
- Identifying the forms of energy included in the problem.
- Drawing free-body diagrams of the bodies.
- Applying appropriate coordinate systems to the diagrams.
- Applying the equations of motion to find relations between the forces, masses, and accelerations.
- Applying conservation of energy or momentum when appropriate.
- Solving the equations through algebra and substitutions.
- Reflecting on the results and checking for inconsistencies.

Expert problem solvers use this foundation at all levels of physics investigations, from introductory courses through cutting-edge research projects.

11 ELASTICITY AND PERIODIC MOTION

Summary

Objects can be stretched, twisted, and compressed. In this chapter, we will examine deformations that describe that stretching, twisting, and compressing. New concepts and principles will be introduced so that we may quantify deformations; these ideas are based on the concepts and principles we encountered in previous chapters. We will delve into one special form of deformation: elastic behavior that leads to periodic motion or oscillation. Periodic motion plays a vital role in many areas of physics, and this chapter will lay the foundation for further studies.

Objectives

- Learn about stress and strain for tension, compression, and shear forces.
- Use Young's, bulk, and shear moduli to predict the changes due to stress.
- Understand periodic motion and the terminology used to describe oscillations.
- Learn to identify and analyze simple harmonic motion.
- Describe energy and motion as a function of time for a particle in simple harmonic motion.
- Understand the simple pendulum, damped and forced oscillations, and resonance.

Concepts and Equations

Term	Description
Stress and Strain	Stress characterizes the strength of a force that stretches, squeezes, or twists an object. Strain is the resulting deformation. Stress and strain are often directly proportional, with the proportionality given by Hooke's law.
Tensile and Compressive Stress and Strain	Tensile and compressive stresses stretch and compress an object, respectively. Tensile stress is the ratio of the perpendicular component of a force to the cross-sectional area to which the force is applied: $$\text{Tensile stress} = \frac{F_\perp}{A}.$$ The SI unit of stress is the pascal (Pa), equal to 1 newton per meter squared. The tensile strain is the ratio of the change in an object's length under stress to its original length: $$\text{Tensile Strain} = \frac{\Delta l}{l_0}.$$ Compressive stress is the result of a pushing force. Compressive strain results in a reduction in an object's length and is generally associated with a negative change in length.
Young's Modulus	Young's modulus (Y) is the ratio of stress to strain: $$Y = \frac{\text{stress}}{\text{strain}} = \frac{l_0 F_\perp}{A\Delta l}.$$ Young's modulus is constant for small stresses. Materials with large Y are less stretchable than materials with small Y.
Pressure and Volume Stress and Strain	Volume stress or pressure is the force per unit area of a fluid: $$p = \frac{F_\perp}{A}.$$ The force is uniform and acts in all directions to increase or decrease an object's volume. Common units of pressure include the pascal and the atmosphere. Volume strain is the fractional change in volume due to pressure: $$\text{Volume strain} = \frac{\Delta V}{V_0}.$$
Bulk Modulus	The bulk modulus B is the ratio of volume stress to strain wherever Hooke's law is valid: $$B = -\frac{\Delta p}{\Delta V/V_0}.$$ The negative sign indicates that an increase in pressure results in a decreased volume.
Shear Stress, Strain, and Modulus	Shear stress is the force tangent to an object's surface divided by the area over which the force acts: $$\text{Shear stress} = \frac{F_\parallel}{A}.$$ Shear strain is the ratio of the displacement x to the transverse dimension h: $$\text{Shear strain} = \frac{x}{h} = \tan\phi.$$

The shear modulus (S) is the ratio of shear stress to strain wherever Hooke's law is valid:

$$S = \frac{F_{\parallel}/A}{\phi}.$$

Elasticity and Plasticity	Elastic deformations are reversible (i.e., a stressed object will return to its original size upon release). Plastic deformations occur when an object returns close to its original size after release.
Periodic Motion	Periodic motion is motion that repeats in a definite cycle. Periodic motion occurs when an object is displaced from its equilibrium position and a restoring force exists that tends to return the object to equilibrium. The amplitude is the maximum magnitude of displacement from equilibrium. A cycle is one complete round-trip. The period is the time taken to complete one cycle. The frequency (f) is the number of cycles per unit time. The SI unit of frequency is the hertz, (Hz), equal to 1 per second. The angular frequency (ω) is 2π times the frequency. Period, frequency, and angular frequency are related: $$T = \frac{1}{f}, \qquad f = \frac{1}{T}, \qquad \omega = 2\pi f = \frac{2\pi}{T}.$$
Simple Harmonic Motion	Simple harmonic motion (SHM) is periodic motion in which the restoring force is directly proportional to an object's displacement. The total mechanical energy remains constant in SHM. The equations of motion are $$x = A\cos\omega t = A\cos(2\pi ft),$$ $$v_x = -\omega A\sin\omega t = -2\pi fA\sin(2\pi ft),$$ $$a_x = -\omega^2 A\cos\omega t = -(2\pi f)^2 A\cos(2\pi ft).$$ A system with spring constant k and mass m has frequency $$f = \frac{\omega}{2\pi} = \frac{1}{2\pi}\sqrt{\frac{k}{m}}$$ and period $$T = \frac{1}{f} = 2\pi\sqrt{\frac{m}{k}}.$$
Simple Pendulum	A simple pendulum is a model of a point mass suspended by a weightless string in a gravitational field. For small displacements, a pendulum of length L has frequency $$f = \frac{\omega}{2\pi} = \frac{1}{2\pi}\sqrt{\frac{g}{L}}$$ and period $$T = \frac{2\pi}{\omega} = \frac{1}{f} = 2\pi\sqrt{\frac{L}{g}}.$$
Damped and Forced Oscillations	Periodic motion in a system that has dissipative forces is called *damped oscillation*. Periodic motion in a system with a periodically changing force is called *forced oscillation* or *driven oscillation*. Resonance occurs when the driving angular frequency is near the natural oscillation angular frequency, increasing the amplitude of the motion.

Conceptual Questions

1: Glider in simple harmonic motion

A glider attached to a spring and set on a horizontal air track is allowed to oscillate with a 5.0 cm amplitude. How far does the glider travel in one period?

Solution

SET UP AND SOLVE We answer the question by considering how the glider moves during one period. The period is the time an object takes to move from any position through one complete cycle and return to the starting position. Starting from the equilibrium position, imagine that the glider moves to the right and momentarily stops at a displacement equal to the amplitude. It has traveled a distance of one amplitude, or 5.0 cm. The glider then returns to its equilibrium position, traveling a second distance equal to the amplitude, or a total of 10.0 cm. The glider continues moving to the left until it reaches its maximum displacement on the left side, thus traveling a third distance equal to the amplitude (15.0 cm total). The glider then returns to the right and to the starting position, traveling a fourth distance equal to the amplitude (20.0 cm total).

In one period, the glider travels a distance equal to four amplitudes, or 20.0 cm.

REFLECT You must distinguish between amplitude and total distance traveled. Comprehending this difference helps build an understanding of simple harmonic motion.

2: Gravity on the moon

You are asked to estimate the moon's gravitational acceleration by watching a video of the early lunar explorations. How could you estimate the acceleration due to gravity on the moon?

Solution

SET UP AND SOLVE We've seen that, for small oscillations, the period of a simple pendulum is related to the gravitational constant and length of the pendulum. If you can find an object that can be approximated by a simple pendulum, then you can determine the moon's gravitational acceleration from the object's motion. One approach would be to look for a dangling object during a moonwalk. You can estimate the length of the pendulum by comparing it with the size of the astronaut on the walk and measure the time with a stopwatch or by counting video frames.

REFLECT The moon's gravitational acceleration was estimated by physics students around the world who watched the early moonwalks. The technique can also be used to estimate the sizes of objects in videos by taking the known gravitational acceleration value and combining it with the period to find the length of the pendulum.

Problems

1: Strain on an elevator cable

A steel elevator cable can support a maximum stress of 9.0×10^7 Pa. If the maximum weight of the fully loaded elevator is 2100 kg and the maximum upward acceleration is 3.0 m/s^2, what should the diameter of the cable be? By how much does the cable stretch when the elevator is accelerating upwards at 3.0 m/s^2, and 120 m of cable has been released? (Young's modulus for steel is 2×10^{11} Pa).

Solution

SET UP For the first part of the problem, we'll use Newton's second law to find the tension in the elevator cable and then use the maximum stress to find the cable diameter. To solve the second part, we'll use Young's modulus.

Figure 11.1 shows the free-body diagram for the elevator. Gravity and tension act on the elevator.

▲ **Figure 11.1**

SOLVE We'll apply Newton's second law to find the maximum tension in the cable:

$$\sum F_y = T - mg = ma_y.$$

The tension is then

$$T = m(g + a_x) = (2100 \text{ kg})((9.8 \text{ m/s}^2) + (3.0 \text{ m/s}^2)) = 26{,}900 \text{ N}.$$

Stress is the force per unit area, or

$$S = \frac{F}{A} = \frac{T}{A}.$$

The area can be written in terms of the diameter as

$$A = \frac{\pi d^2}{4}.$$

The diameter is then

$$d = 2\sqrt{\frac{T}{S\pi}} = 2\sqrt{\frac{(26{,}900 \text{ N})}{(9.0 \times 10^7)\pi}} = 0.020 \text{ m} = 2.0 \text{ cm},$$

where we have replaced the stress with the maximum stress. Young's modulus leads to the amount of cable stretch:

$$Y = \frac{l_0 F}{A \Delta l} = \frac{4 l_0 T}{\pi d^2 \Delta l}.$$

Rearranging terms to find the cable stretch gives

$$\Delta l = \frac{4 l_0 T}{\pi d^2 Y} = \frac{4(120 \text{ m})(26{,}900 \text{ N})}{\pi (0.020 \text{ m})^2 (2.0 \times 10^{11} \text{ Pa})} = 0.051 \text{ m} = 5.1 \text{ cm}.$$

The cable must have a 2.0 cm diameter and stretches 5.1 cm when 120 m of cable has been released.

REFLECT We see that the 2-cm-thick cable stretches over 5 cm. This may appear to be a significant elongation, but it represents only 0.04% of the cable's length.

2: Mass on a spring

A spring with a 1.2 kg mass hung from it stretches 4.7 cm from its equilibrium position. If the mass is now stretched 6.5 cm from its equilibrium position and released, find (a) the period of the motion, (b) the maximum velocity, and (c) the maximum acceleration.

Solution

SET UP We'll use the equations of simple harmonic motion to find the solutions to the problem. We first find the spring constant, using the preliminary information.

SOLVE We find the spring constant from Hooke's law. When the mass is initially attached to the spring, it hangs in equilibrium, with the spring force equal to gravity:

$$F_s = kx = mg.$$

The spring constant is

$$k = \frac{mg}{x} = \frac{(1.2 \text{ kg})(9.8 \text{ m/s}^2)}{0.047 \text{ m}} = 250 \text{ N/m}.$$

With the spring constant, we can directly find the period:

$$T = 2\pi\sqrt{\frac{m}{k}} = 2\pi\sqrt{\frac{(1.2 \text{ kg})}{(250 \text{ N/m})}} = 0.44 \text{ s}.$$

The maximum velocity is

$$v_{max} = \sqrt{\frac{k}{m}}A = \sqrt{\frac{(250 \text{ N/m})}{(1.2 \text{ kg})}}(0.065 \text{ m}) = 0.94 \text{ m/s}.$$

The maximum (positive) acceleration occurs when the mass is at its most negative position, so

$$a_{max} = -\frac{k}{m}x = -\frac{(250 \text{ N/m})}{(1.2 \text{ kg})}(-0.065 \text{ m}) = 13.5 \text{ m/s}^2.$$

The mass oscillates with a period of 0.44 s and has a maximum velocity of 0.94 m/s and a maximum acceleration of 13.5 m/s².

REFLECT Simple harmonic motion is the most complicated motion we have studied to date. However, our previous experiences led to straightforward relations, from which we can now easily extract useful information.

3: Period of a simple pendulum

A simple pendulum reaches a maximum angle of 7.2° after swinging through the bottom of its path with a maximum speed of 0.35 m/s. What is the period of the pendulum's oscillation?

Solution

SET UP We'll use the maximum angle to find the pendulum's amplitude. The amplitude and the maximum speed will give the length of the pendulum, and the period will be derived from the length.

SOLVE The amplitude of a simple pendulum is the maximum arc length, which is related to the maximum angle by the length:

$$S_{max} = L\theta_{max}.$$

The maximum velocity is

$$v_{max} = 2\pi f A = 2\pi f L\theta_{max}.$$

For a simple pendulum, the frequency is given by the length:

$$f = \frac{1}{2\pi}\sqrt{\frac{g}{L}}.$$

Combining these formulas, we find the length:

$$v_{max} = 2\pi\left(\frac{1}{2\pi}\sqrt{\frac{g}{L}}\right)L\theta_{max} = \sqrt{gL}\,\theta_{max},$$

$$L = \frac{v_{max}^2}{g\theta_{max}^2} = \frac{(0.35 \text{ m/s})^2}{(9.8 \text{ m/s}^2)(0.126)^2} = 0.79 \text{ m}.$$

Note that we replaced the maximum angle of 7.2° with the equivalent 0.126 radian. The period is then

$$T = 2\pi\sqrt{\frac{L}{g}} = 2\pi\sqrt{\frac{(0.79 \text{ m})}{(9.8 \text{ m/s}^2)}} = 1.8 \text{ s}.$$

The pendulum has length 79 cm and period 1.8 s.

REFLECT The simple pendulum's period depends only on length and the gravitational constant. The maximum angle and velocity provided enough information to solve the problem.

4: Oscillating blocks

Two blocks shown in Figure 11.2 oscillate on a frictionless surface with a frequency of 0.30 Hz. The top block has mass 2.0 kg and the bottom block has mass 4.5 kg. If the amplitude is increased to 25 cm, the top block begins to slide. What is the coefficient of static friction?

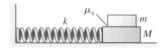

▲ **Figure 11.2**

Solution

SET UP When the top block just begins to slide, the force applied must be equal to the maximum static frictional force. The maximum applied force occurs at the maximum displacement (equal to the amplitude). We'll need the spring constant, which we extract from the initial frequency. The motion is simple harmonic, as the only horizontal force acting on the blocks is the spring force, a restoring force that is directly proportional to the displacement.

SOLVE The frequency in simple harmonic motion depends on the spring constant according to the formula

$$f = \frac{1}{2\pi}\sqrt{\frac{k}{m + M}},$$

where we include the combined mass of the oscillating blocks. Solving for k gives

$$k = (2\pi f)^2(m + M) = (2\pi(0.30 \text{ Hz}))^2((2.0 \text{ kg}) + (4.5 \text{ kg})) = 23.1 \text{ N/m}.$$

Recall that the maximum static frictional force is $f_s = \mu_s n$. For the top block, the normal force is mg. The maximum force applied by the spring is kA. Equating these forces yields

$$f_s = \mu_s n = \mu_s mg = kA.$$

Rearranging terms to find the coefficient of static friction results in

$$\mu_s = \frac{kA}{mg} = \frac{(23.1 \text{ N/m})(0.25 \text{ m})}{(2.0 \text{ kg})(9.8 \text{ m/s}^2)} = 0.29.$$

The coefficient of static friction between the blocks is 0.29.

REFLECT This problem brings together topics from several areas we have studied throughout the text, including the normal force, frictional forces, the spring force, and simple harmonic motion. Combining our knowledge helps us understand complex phenomena.

12 MECHANICAL WAVES AND SOUND

Summary

In this chapter, we expand the concept of the periodic motion of an object to the periodic motion of many particles connected together as a medium. The periodic motion of a medium is a mechanical wave. Waves occur in many forms, including ocean waves, sound, light, earthquakes, and television transmission. This chapter will form the foundation for studying a variety of waves. We'll begin with the description of transverse and longitudinal waves and their amplitudes, periods, frequencies, and wavelengths. We'll see how waves move, interact, reflect, and combine in a variety of ways, and we'll learn how to describe their frequencies. We'll examine how the frequency of a wave changes relative to how the source and listener are moving, which is summarized in the Doppler effect.

Objectives

- Identify longitudinal and transverse waves and their media.
- Learn the relations between the period, velocity, frequency, and wavelength of a wave.
- Understand the concepts of superposition, standing waves, nodes, and antinodes.
- Learn the allowed frequencies for transverse and longitudinal standing waves.
- Learn about the phenomena of interference for interacting waves.
- Define and use sound-wave intensities and beats.
- Apply the Doppler effect to moving sources and listeners.
- Survey the application of acoustics to a variety of systems.

Concepts and Equations

Term	Description
Mechanical Wave	A mechanical wave is a disturbance from equilibrium that propagates from one region of space to another through a medium. In a transverse wave, the particles in the medium are displaced perpendicular to the direction of travel. In a longitudinal wave, the particles in the medium are displaced parallel to the direction of travel.
Periodic Mechanical Waves	In a periodic mechanical wave, particles in the medium exhibit periodic motion. The speed, wavelength, period, and frequency of a periodic wave are related by $$v = \lambda f = \frac{\lambda}{T}.$$ The speed of a transverse wave in a rope under tension is given by $$v = \sqrt{\frac{F_T}{\mu}},$$ where F_T is the tension in the rope and μ is the mass per length.
Principle of Superposition	The principle of superposition states that when two waves overlap, the net displacement at any point at any time is found by taking the vector sum of the position given by the two individual waves.
Standing Waves	A standing wave is the combination of waves that produces a stationary pattern. Nodes are points at which the standing wave pattern does not change with time. Antinodes are positions halfway between nodes; the amplitude is maximum at an antinode. The distance between successive nodes or antinodes is one-half of the wavelength. A string of length L held stationary at both ends can have only frequencies such that $$f_n = n\frac{v}{2L} = nf_1 \qquad (n = 1, 2, 3, \dots).$$ f_1 is the fundamental frequency; the multiples of f_1 are the harmonics.
Longitudinal Standing Waves	Longitudinal waves that propagate in a fluid in a pipe can reflect and form longitudinal standing waves. For a pipe with an open end, the fundamental frequency and harmonics are $$f_n = n\frac{v}{2L} = nf_1 \qquad (n = 1, 2, 3, \dots).$$ For a pipe with a closed end, the fundamental frequency and harmonics are $$f_n = n\frac{v}{4L} = nf_1 \qquad (n = 1, 3, 5, \dots).$$
Interference	When waves overlap in the same region of space, the waves are said to interfere. When the waves combine to form a wave with a larger amplitude, the waves constructively interfere or reinforce each other. When the waves differ by a half cycle, their sum results in a wave with a smaller amplitude, and the waves destructively interfere, or cancel.
Sound	Sound is a pressure wave that travels through gas, liquid, and solids. The intensity level β of a sound wave is $$\beta = (10\,\text{dB})\log\frac{I}{I_0},$$ where I_0 is the reference intensity $(10^{-12}\,\text{W/m}^2)$. The units of β are decibels (dB).

Doppler Effect	The Doppler effect is the frequency shift that occurs when the listener is in motion relative to the source of sound. The listener's frequency f_L is related to the source frequency f_S by $$f_L = \frac{v + v_L}{v + v_S} f_S,$$ where v is the speed of sound and v_L and v_S are the x components of the speed of the listener and source, respectively.

Conceptual Questions

1: Waves in a jump rope

Your younger sister is playing with a jump rope. She ties one end to a fence post and moves the other end up and down, observing waves in the rope. She sees waves moving down the rope and becomes confused. She asks you why the rope doesn't move towards the fence, since that is how the waves move. How do you answer her question?

Solution

SET UP AND SOLVE Having just learned about mechanical waves, you explain that the rope is made up of lots of little pieces (particles) and that the pieces at the end she holds move up and down as she moves her hand up and down. The pieces of rope next to her end are connected to the pieces she is moving, so are pulled up and down at the same time. These pieces, which take a little bit longer to move than the ones she holds, are connected to other pieces, which also move up and down. Each successive piece takes a bit longer to start moving than the piece before it, which is why there is a wave pattern. This wave pattern is what your sister observes moving down the rope. All of the individual pieces of rope move only up and down. Since they are connected to each other, their moving creates a disturbance in the rope that appears to move down the rope. The rope doesn't move towards the fence, since none of the pieces of the rope move towards the fence.

REFLECT Remember that when you pluck a string, you pull the string to the side, so you give the string a velocity to the side, or perpendicular to the string. You impart a force *perpendicular* to the string during the pluck, not along the string.

2: Explaining Doppler shift

Your younger brother asks you to explain why the sound from train whistles changes from a high pitch to a low pitch when a train passes. How do you explain the change?

Solution

SET UP AND SOLVE You first explain that sound comes from vibrations or changing pressure. A high pitch comes from faster vibrations, a low pitch from slower vibrations. The train whistle produces a steady number of vibrations. When the train approaches, the vibrations become compressed, effectively increasing their number. When the train leaves, the vibrations become expanded, effectively slowing them down.

REFLECT This description provides an alternative explanation of the Doppler shift. As the listener and source move towards each other, the wave fronts become closer together and the frequency increases.

3: An orchestra warming up

While you wait for an orchestra to perform, you hear the musicians tuning their instruments. You observe that several will play the same note for several seconds to check the tune. What are they doing?

Solution

SET UP AND SOLVE The musicians are trying to play the same frequency when they tune their instruments. If one or more instruments vibrate at a slightly different frequency, the different frequencies interfere and produce beats. When tuning, the musicians listen for beats and readjust their instruments until the beats are removed.

REFLECT You can listen for beats when you hear music. Some beats are intentional; some pipe organs have a slow beat to create an undulating effect.

Problems

1: An unusual scale

Your strange physics professor builds an unusual scale by hanging an object from a 3.0-m-long wire attached to the ceiling. She plucks the string just above the object and finds that the pulse takes 0.50 s to propagate up and down the wire. What is the mass of the object? The mass of the wire is 0.50 kg.

Solution

SET UP The speed of a wave in a wire under tension is related to the tension in the wire. By finding the speed of the wave in the wire, we'll determine the mass of the object.

SOLVE The speed of the wave in the wire is given by

$$v = \sqrt{\frac{F_T}{\mu}}.$$

The tension at the bottom of the wire is equal to the gravitational force on the object, since the object is in equilibrium. To calculate the tension, we first need the velocity and mass per unit length. The velocity is found by noting that the wave takes 0.50 s to travel 6.0 m (up and down the wire). Thus,

$$v = \frac{\Delta d}{\Delta t} = \frac{6.0 \text{ m}}{0.5 \text{ s}} = 12 \text{ m/s}.$$

The mass per unit length is

$$\mu = \frac{m}{L} = \frac{0.50 \text{ kg}}{3.0 \text{ m}} = 0.167 \text{ kg/m}.$$

Combining terms to find the tension, we obtain

$$F_T = v^2\mu = (12 \text{ m/s})^2(0.167 \text{ kg/m}) = 24 \text{ N}.$$

The tension force is equal to the weight, so the mass of the object is

$$m = \frac{F_T}{g} = \frac{24 \text{ N}}{9.8 \text{ m/s}^2} = 2.4 \text{ kg}.$$

The object's mass is 2.4 kg.

REFLECT This unusual scale illustrates how we can use mechanical waves to measure mass, but it is impractical for several reasons. First, the scale requires a high ceiling and a method of accurately measuring the speed of waves in the wire. Also, we have omitted the mass of the wire, which is roughly 15% higher at the ceiling, in the calculation of the tension.

2: Overtones in a string

A steel wire 1.3 m long having a mass of 12 grams is under 750 N of tension. How many overtones can be heard by a person with good hearing?

Solution

SET UP The number of overtones is the number of frequencies about the fundamental frequency. A person with good hearing can hear in the range from 20 Hz to 20,000 Hz. We'll find the number of frequencies in that range for the string and subtract the fundamental frequency to solve the problem.

SOLVE The frequencies in a standing wave in a wire are

$$f_n = nf_1,$$

where n is an integer. The fundamental frequency for a wire is

$$f_1 = \frac{1}{2L}\sqrt{\frac{F_T}{\mu}}.$$

For this wire,

$$f_1 = \frac{1}{2L}\sqrt{\frac{F_T}{\mu}} = \frac{1}{2(1.3 \text{ m})}\sqrt{\frac{(750 \text{ N})}{0.012 \text{ kg}/1.3 \text{ m}}} = 109.6 \text{ Hz}.$$

The fundamental frequency is above 20 Hz, so it can be heard. The highest frequency that can be heard is 20,000 Hz, which corresponds to

$$n = \frac{20,000 \text{ Hz}}{f_1} = \frac{20,000 \text{ Hz}}{109.6 \text{ Hz}} = 182.4.$$

Since we cannot hear four-tenths of a frequency, we truncate n to 182. Thus, 182 frequencies can be heard: the fundamental frequency and 181 overtones.

REFLECT This problem illustrates how to count overtones carefully and how to interpret integer results. We'll see more examples of these techniques as we progress through the text.

3: Power at a concert

You are given the task of determining how much power is needed in the sound system at a new stadium. If the design calls for an intensity of 100 dB at the furthest seats (120 m from the speakers), how much power is required?

Solution

SET UP We assume that the sound is distributed over a sphere of radius 120 m in order to determine the intensity from the power. We'll use the definition of intensity level to relate the design intensity to the power.

SOLVE The intensity level is given by

$$\beta = (10 \text{ dB}) \log \frac{I}{I_0}.$$

Taking the logarithm of both sides gives

$$I = I_0 10^{(\beta/10 \text{ dB})}.$$

The intensity is the power per unit area; in this case, the area is that of a sphere $(4\pi r^2)$. Combining the various equations gives

$$P = AI = (4\pi r^2)(I_0 10^{(\beta/10 \text{ dB})}) = (4\pi(120 \text{ m})^2)((10^{-12} \text{ W/m}^2)10^{((120 \text{ dB})/10 \text{ dB})}) = 180 \text{ kW}.$$

The required power is 180 kW.

REFLECT Our stadium requires a substantial sound system for all visitors to hear the concert. This amount of power would harm the hearing of those near the speaker. Stadiums are designed with multiple speakers placed at strategic locations and closer to the visitors in order to reduce the maximum volume.

Practice Problem: What is the intensity for persons seated 20 m from the speakers? *Answer:* 135 dB, above the threshold for permanent hearing damage.

4: Speed of approaching train

You are driving along a country road at 20.0 m/s. A train approaches on a rail that parallels the road. The train whistle blasts at a frequency of 800 Hz, but you hear a 950 Hz whistle. What is the speed of the approaching train?

Solution

SET UP The Doppler effect describes the frequency shift for moving sources, so we'll use the Doppler formula to determine the speed of the approaching train. Our coordinate system is shown in Figure 12.1; positive velocities are taken to be from the listener towards the source. Both the source and listener are moving. The listener's velocity is positive and the source's velocity is negative in our coordinate system. The speed of sound is 340 m/s.

▲ **Figure 12.1**

SOLVE The listener's frequency is related to the source frequency by the Doppler shift

$$f_L = \frac{v + v_L}{v + v_S} f_S,$$

where v is the speed of sound and v_L and v_S are the x components of the speed of the listener and source, respectively. We can rearrange terms to solve for v_S:

$$v_S = \frac{v + v_L}{f_L f_S - v}.$$

In this case v_L is $+20.0$ m/s, f_L is 950 Hz, f_S is 800 Hz, and v is 340 m/s. Substituting yields

$$v_S = \frac{v + v_L}{f_L} f_S - v = \frac{(340 \text{ m/s}) + (20.0 \text{ m/s})}{(950 \text{ Hz})}(800 \text{ Hz}) - (340 \text{ m/s}) = -36.8 \text{ m/s}.$$

The train is approaching at 36.8 m/s, or 132 kilometers per hour.

REFLECT This problem illustrates how to use the Doppler shift to find the speed of an object. We expected and found a negative speed, indicating that the source was moving towards the listener. We see that proper Doppler-shift solutions require a coordinate system and careful interpretation of the velocity's directions.

Practice Problem: What frequency would you hear if you were moving away from the train at 20.0 m/s? *Answer:* 844 Hz.

13 FLUID MECHANICS

Summary

We interact with fluids on a continual basis, from walking through air to swimming in the ocean. This chapter examines fluids, or substances that can flow, including liquids and gases. We will begin with fluid statics and use Newton's laws to describe the behavior of fluids at rest. Density, pressure, buoyancy, and surface tension are concepts needed for our investigation, and they will be defined. We will also delve into fluid dynamics and see how to analyze fluids in motion. Conservation of energy and Newton's laws will guide us in this examination. While fluid dynamics can be quite complex, several examples will give us insight into the subject.

Objectives

- Define fluids and analyze their properties.
- Learn about and understand density, pressure, buoyancy, and surface tension.
- Analyze fluids in equilibrium and find the pressure at varying depths.
- Understand and apply the buoyant force to problems involving fluids.
- Apply Bernoulli's equation to problems in fluid dynamics.

Concepts and Equations

Term	Description
Density	Density is the mass per volume of a material. For a material with mass m and volume V, the density is $$\rho = \frac{m}{V}.$$ The SI unit of density is the kilogram per cubic meter $(1\ \text{kg/m}^3)$. The cgs unit used to express density is the gram per cubic centimeter $(1\ \text{gm/cm}^3 = 1000\ \text{kg/m}^3)$.
Specific Gravity	The specific gravity of a material is the ratio of the density of the material to that of water.
Pressure	The pressure p in a fluid is the magnitude of the perpendicular force $F_\perp$ exerted on a plane surface area A, divided by the area: $$p = \frac{F_\perp}{A}.$$ The SI unit of pressure is the pascal (Pa); $1\ \text{Pa} = 1\ \text{N/m}^2$. Also common are the bar $(10^5\ \text{Pa})$ and millibar $(10^2\ \text{bar})$.
Pressure in a Fluid	The pressure p in a fluid at a height h relative to a reference level $h = 0$ and reference pressure p_0 is $$p - p_0 = -\rho g h,$$ where ρ is the density of the fluid. Pascal's law states that the pressure applied to a fluid is transmitted through the fluid and depends only on depth.
Buoyant Force	Archimedes' principle states that when an object is immersed in a fluid, the fluid exerts an upward buoyant force on the object equal in magnitude to the weight of the fluid displaced by the object.
Surface Tension	Surface tension γ is the ratio of the magnitude F of the surface force to the length d along which the force acts: $$\gamma = \frac{F}{d}.$$ The SI unit of surface tension is the newton per meter (N/m).
Fluid Flow	An ideal fluid is incompressible and has no viscosity. Conservation of mass requires that the amount of fluid flowing through a cross section of a tube per unit time be the same for all cross sections: $$\frac{\Delta V}{\Delta t} = A_1 v_1 = A_2 v_2.$$
Bernoulli's Equation	Bernoulli's equation relates the pressure p, flow speed v, and elevation y for a ideal fluid at any two points: $$p_1 + \rho g y_1 + \tfrac{1}{2}\rho v_1^2 = p_2 + \rho g y_2 + \tfrac{1}{2}\rho v_2^2 = \text{constant}.$$

Conceptual Questions

1: Ice in a glass

Two glasses are filled with water to the same level. In one glass, ice cubes float on the top. If the two glasses are made of the same material and have the same shape, how does their total weight compare?

Solution

SET UP AND SOLVE The glasses must have the same weight, since they are made of the same material and have the same shape. We solve the problem by comparing the mass of the water alone to the mass of the water-plus-ice mix.

Archimedes' principle states that an object will displace its own weight in a fluid. The volume of water displaced by the ice has the same weight as the ice; therefore, the water in one glass weighs the same as the water plus ice in the second glass. The total combined weight is the same.

REFLECT The volume of the glass with the ice is greater than that of the glass without ice, but weight depends on both density and volume. As with any new physical principle, we need to develop our skills carefully and not jump to conclusions.

2: Energy in a hydraulic lift

A hydraulic lift is used to lift a car. The piston supporting the car has a cross-sectional area 100 times larger than the cross-sectional area of the piston driving the lift. The drive piston will therefore require a force 100 times smaller than the weight of the car to lift the car. Does this mean that energy conservation is violated?

Solution

SET UP AND SOLVE Pascal's law states that the pressure is the same at both pistons. The drive piston requires a small force to create the pressure that will lift the car. A large displacement in the drive piston creates a small displacement in the lift piston, due to the differences in areas. The amount of work done in moving the drive piston a long distance is equal to the work done by the lift piston in moving a small distance (ignoring friction). The work is equivalent; energy conservation is not violated.

REFLECT If you have ever operated a hydraulic jack to lift you car or a house, you should recall that you had to pump the jack several times to move a small distance. The work from your small force applied over a long distance was equivalent to the work done in lifting the object.

3: Race-car spoilers

Why do race cars have spoilers, or wings, on their bodies?

Solution

SET UP AND SOLVE Spoilers are essentially inverted airplane wings. We've seen that airplane wings produce lift for the planes by reducing the pressure above the wing. The inverted wing produces a *downward* force to help hold the race car on the pavement and maintain contact between the wheels and the road. The spoiler also helps stabilize the car as it moves around the track.

REFLECT Spoiler design for race cars is critical; there must be a careful balance between enough downward force to keep the car on the track and too much force, which causes lost fuel economy and premature tire wear. Some race cars have downward forces of up to three times the force of gravity, (i.e., they could operate on an upside-down track and not fall off). Many cars have spoilers, though most have only an aesthetic value and no effect on the car's performance.

Problems

1: How much seawater in a tank?

Seawater is stored under pressure in a tank of horizontal cross-sectional area 4.5 m^2. The pressure above the seawater in the tank is 7.2×10^5 Pa, and the pressure at the bottom of the tank is 1.2×10^6 Pa. What is the mass of the seawater in the tank?

Solution

SET UP The tank is sketched in Figure 13.1. We can find the mass by first finding the volume of the seawater and then multiplying by the density. To find the volume, we multiply the height by the cross-sectional area of the tank. To find the height, we use the variation in the pressures due to the height of seawater. We assume that the seawater is incompressible.

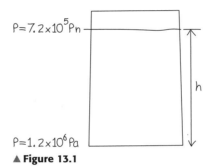

P=7.2×10^{5}Pn

h

P=1.2×10^6 Pa

▲ **Figure 13.1**

SOLVE We start by finding the height of the seawater in the container. The difference in pressure is related to the height by

$$p - p_0 = \rho g h,$$

where we set p_0 as the pressure at the bottom of the tank and p as the pressure at the top of the tank. Solving for h, we get

$$h = \frac{p - p_0}{\rho g} = \frac{(1.2 \times 10^6 \text{ Pa}) - (7.2 \times 10^5 \text{ Pa})}{(1.03 \times 10^3 \text{ kg/m}^3)(9.8 \text{ m/s}^2)} = 47.6 \text{ m},$$

where we used the density of seawater given in Table 13.1 (1.03×10^3 kg/m^3). We now find the volume of seawater in the tank. The volume is the height times the cross-sectional area:

$$V = hA = (47.6 \text{ m})(4.5 \text{ m}^2) = 214 \text{ m}^3.$$

The mass is the volume times the density we used from the table:

$$m = \rho V = (1.03 \times 10^3 \text{ kg/m}^3)(214 \text{ m}^3) = 221,000 \text{ kg}.$$

The tank holds 221,000 kg of seawater.

REFLECT This problem shows how to determine height from change in pressure. Altimeters find changes in altitude by monitoring the change in pressure of air.

2: Velocity of water out of a fire hose

Water enters a round fire hose of diameter 3.5 cm and exits out of a round, 0.60-cm-diameter nozzle. If the water enters the hose at 2.0 m/s, what is the velocity of the exiting water? What is the maximum horizontal range of the water leaving the hose?

Solution

SET UP The continuity equation relates the velocities and cross-sectional areas for incompressible fluids in a tube. We can use that equation to find the velocity of the water exiting the nozzle. To find the range, we use results from projectile motion. We treat the water as incompressible.

SOLVE The amount of fluid flowing through a tube per time is constant. The flow through the hose is equal to the flow through the nozzle:

$$A_{\text{hose}} v_{\text{hose}} = A_{\text{nozzle}} v_{\text{nozzle}}.$$

The area of the hose or nozzle is π times the square of half the diameter. Solving for the velocity of the nozzle gives

$$v_{\text{nozzle}} = \frac{A_{\text{hose}} v_{\text{hose}}}{A_{\text{nozzle}}} = \frac{\pi (D_{\text{hose}}/2)^2 v_{\text{hose}}}{\pi (D_{\text{nozzle}}/2)^2} = \frac{(3.5 \text{ cm})^2 (2.0 \text{ m/s})}{(0.6 \text{ cm})^2} = 68 \text{ m/s},$$

where the factors π and 2 cancel. The water molecules leaving the hose have a velocity of 68 m/s and undergo acceleration due to gravity. We can use the kinematic relations for two-dimensional motion to find the range. Recall from Example 3.5 that the horizontal range of a projectile in terms of the launch angle θ_0 and initial velocity v_0 is

$$R = \frac{v_0^2 \sin 2\theta_0}{g}.$$

The maximum range occurs when then launch angle is 45°. Substituting our values, we obtain

$$R = \frac{v_0^2 \sin 2(45°)}{g} = \frac{(68 \text{ m/s})^2 (1)}{(9.8 \text{ m/s}^2)} = 470 \text{ m}.$$

The maximum horizontal range for the water leaving the nozzle at 68 m/s is 470 m.

REFLECT This problem illustrates why nozzles are placed at the ends of hoses. The reduced diameter of the nozzle increases the exiting velocity, and thereby the range, of the water. Next time you wash your car, compare the velocity and range of the water leaving the hose with and without the nozzle attached.

3: Examining the buoyant force

A wooden sphere with a 4.5 cm radius is held in fresh water below the surface by a spring. If the spring's force constant is 55 N/m, by how much is the spring stretched from its equilibrium position? Take the density of wood to be 700 kg/m^3.

Solution

SET UP The free-body diagram for the block of wood is shown in Figure 13.2. The buoyant force is upward, and both gravity and the tension due to the spring are downward. The block of wood is in equilibrium, so the forces sum to zero. The size and density of the wood determine its volume and mass, needed to calculate the buoyant force and gravity. Hooke's law will be used to determine the amount of stretch.

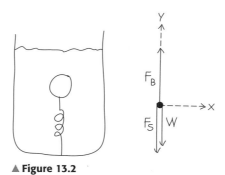

▲ **Figure 13.2**

SOLVE We sum the three forces acting on the block of wood. The forces act only in the vertical direction and add to zero:

$$\sum F_y = F_\text{B} - mg - F_\text{s} = 0.$$

The buoyant force is equal to the amount of water displaced by the wooden sphere. The volume of a sphere is $4/3\pi r^2$. Combining terms yields

$$F_\text{B} = \rho_\text{water} V_\text{sphere} g = \rho_\text{water}\left(\tfrac{4}{3}\pi r_\text{sphere}^3\right) g.$$

The spring force is equal to the spring constant times the displacement kx of the spring. The mass of the sphere is its volume times its density. Substituting into the equilibrium equation gives

$$\sum F_y = F_\text{B} - mg - F_\text{s} = \rho_\text{water}\left(\tfrac{4}{3}\pi r_\text{sphere}^3\right) g - \rho_\text{wood}\left(\tfrac{4}{3}\pi r_\text{sphere}^3\right) g - kx = 0.$$

Rearranging terms enables us to find x:

$$x = \frac{\rho_\text{water}\left(\tfrac{4}{3}\pi r_\text{sphere}^3\right) g - \rho_\text{wood}\left(\tfrac{4}{3}\pi r_\text{sphere}^3\right) g}{k} = \frac{\left(\rho_\text{water} - \rho_\text{wood}\right)\left(\tfrac{4}{3}\pi r_\text{sphere}^3\right) g}{k},$$

$$x = \frac{\left(\left(1 \times 10^3 \text{ kg/m}^3\right) - \left(700 \text{ kg/m}^3\right)\right)\left(\tfrac{4}{3}\pi\left(0.045 \text{ m}\right)^3\right)\left(9.8 \text{ m/s}^2\right)}{\left(55 \text{ N/m}\right)} = 0.020 \text{ m}.$$

The spring is stretched 0.020 m, or 2.0 cm, from its equilibrium position.

REFLECT This problem illustrates how to incorporate the buoyant force with previously encountered forces to solve equilibrium problems. We used the same procedure as in the past, starting with a free-body diagram and setting the net force equal to zero. Working with fluids requires conversions between volume, mass, and density.

4: Lift on a car on a highway

As a car travels down the highway, the speed of the air flowing over the top of the car is higher than the speed of the air flowing under the car, thus creating lift. Estimate the lift on a car as it travels at 100 kph. Take the density of air to be 1.20 kg/m³, the car's area to be 6 m², and the height of the car to be 1.0 m. Assume that the air travels under the car at 100 kph and over the top of the car at 140 kph.

Solution

SET UP Bernoulli's equation gives the pressure difference between the top and bottom of the car. We can find the force from the definition of pressure as force per surface area.

SOLVE The car is moving through a fluid (air), so Bernoulli's equation can be applied. We'll compare the pressures below and above the car to find the pressure difference. Bernoulli's equation is

$$p_{\text{above}} + \rho g y_{\text{above}} + \tfrac{1}{2}\rho v_{\text{above}}^{2} = p_{\text{below}} + \rho g y_{\text{below}} + \tfrac{1}{2}\rho v_{\text{below}}^{2}.$$

The pressure difference is then

$$\Delta p = \rho g y_{\text{above}} + \tfrac{1}{2}\rho v_{\text{above}}^{2} - \tfrac{1}{2}\rho v_{\text{below}}^{2},$$

where we have set the origin below the car $(y_{\text{below}} = 0)$. The pressure difference is

$$
\begin{aligned}
\Delta p &= \rho\left(g y_{\text{above}} + \tfrac{1}{2}\left(v_{\text{above}}^{2} - v_{\text{below}}^{2}\right)\right) \\
&= (1.20 \text{ kg/m}^3)\left((9.8 \text{ m/s}^2)(1.0 \text{ m}) + \tfrac{1}{2}\left((38.9 \text{ m/s})^2 - (27.7 \text{ m/s})^2\right)\right) \\
&= 459 \text{ Pa},
\end{aligned}
$$

where we replaced 100 kpm with 27.7 m/s and 140 kph with 38.9 m/s. We find the force by multiplying the pressure by the area of the car:

$$F = PA = (459 \text{ Pa})(6.0 \text{ m}^2) = 2{,}750 \text{ N}.$$

The lift on the car is 2750 N.

REFLECT We see that the lift is significant in this case—roughly equivalent to a weight of 280 kg. It is not enough to lift the car off the highway, since most cars weigh over 1000 kg. We assumed that the flow of air around the car was smooth and that air is incompressible. Neither are valid assumptions and should be included in a careful examination. Our results show the maximum lift for the car.

14 TEMPERATURE AND HEAT

Summary

In this chapter, we begin a three-chapter investigation into thermodynamics. We will lay the groundwork for the upcoming chapters with an initial definition of temperature and then see how materials change size with temperature. Heat will be introduced as a method of energy transfer due to temperature differences, and the rate of heat transfer will be calculated. We will also learn about the heat required to change the phase of matter and the three types of heat transfer: conduction, convection, and radiation.

Objectives

- Define temperature and thermal equilibrium.
- Learn the three temperature scales.
- Use thermal expansion to find the change in length and volume of materials due to temperature changes.
- Learn of heat, phase changes, and calorimetry and apply these concepts to problems.
- Learn the three forms of heat transfer and apply the heat transfer equation to problems.

Concepts and Equations

Term	Description
Thermal Equilibrium	Two objects in thermal equilibrium have the same temperature.
Temperature Scales	The Celsius temperature scale defines 0°C as the freezing point of water and 100°C as the boiling point of water. The Fahrenheit temperature scale defines 32°F as the freezing point of water and 212° F as the boiling point of water. The Kelvin scale defines absolute zero as 0 K and uses the Celsius unit as its standard unit. Note that 0 K is −273.15°C.
Thermal Expansion	Materials change size, or expand thermally, due to changes in temperature. An object of length L_0 at temperature T_0 will have length L at temperature $T = T_0 + \Delta T$, or $$L = L_0 + \Delta L = L_0(1 + \alpha \Delta T),$$ where α is the coefficient of linear expansion with units K^{-1}. An object of volume V_0 at temperature T_0 will have volume V at temperature $T = T_0 + \Delta T$, or $$V = V_0 + \Delta V = V_0(1 + \beta \Delta T),$$ where β is the coefficient of volume expansion with units K^{-1}.
Heat	Heat is the energy transferred from one object to another due to changes in temperature. The quantity of heat Q needed to raise the temperature of a mass m of material by an amount ΔT is $$Q = mc\Delta T,$$ where c is the specific heat capacity of the material. The SI unit of heat capacity is the joule per kilogram per kelvin (J/(kg K)).
Phase Change	A phase transition is the change from one phase of matter to another. Phases include solid, liquid, and gas. The heat of fusion, L_f, is the heat per unit mass required to change a solid material to liquid. The heat of vaporization, L_v, is the heat per unit mass required to change a liquid material to gas. The heat of sublimation, L_s, is the heat per unit mass required to change a solid material to gas.
Calorimetry	Calorimetry is the measurement of heat in a system. For an isolated system, the algebraic sum of the quantities of heat must add to zero: $$\sum Q = 0.$$
Heat Transfer	Heat may be transferred through conduction, convection, and radiation. Conduction is the transfer of energy within materials without bulk motion of the material. Convection is the transfer of energy due to the motion of mass from one region to another. Radiation is the transfer of energy through electromagnetic radiation. The heat current H for an area A and length L through which the heat flows is given by $$H = \frac{\Delta Q}{\Delta t} = kA\frac{T_H - T_C}{L},$$ where T_H and T_C are, respectively, the temperatures of the hot and cold sides of the material and k is the thermal conductivity. The heat current H due to radiation is $$H = \frac{\Delta Q}{\Delta t} = Ae\sigma T^4,$$ where A is the surface area, e is the emissivity of the surface (a pure number between 0 and 1), T is the absolute temperature, and σ is the Stefan–Boltzman constant $(5.6705 \times 10^{-8} \ W/m^2/K^4)$.

Conceptual Questions

1: Do holes expand or contract?

Your younger brother knows that solids expand as they heat up. He thinks that the metal surrounding the hole in a cookie sheet will expand into the hole as the sheet heats up. Is he right or wrong?

Solution

SET UP AND SOLVE Figure 14.1 shows a sketch of a cookie sheet with a hole in it. After considering the problem, you realize that the hole will enlarge as the sheet heats up, since all dimensions of an object enlarge with temperature. The challenge is how to best explain this to him.

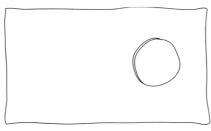

▲ **Figure 14.1**

If you give it a bit more thought, you may come up with a convincing argument. If the cookie sheet had no hole, the whole sheet would increase in size with temperature. If you punch out a hole in the sheet and consider the piece of metal that was removed, this piece expands as its temperature rises. Therefore, the hole in the cookie sheet must also expand, just as it did when the cookie sheet was holeless.

REFLECT Thermal expansion must be carefully considered. Here, we see that a confusing point can be clarified by imagining what happens to the piece that was once the hole.

2: Cooler after a shower

When you step out of the shower, you often feel cold. After drying off, you feel warmer, even though the room's temperature is the same as when you stepped out of the shower. Why?

Solution

SET UP AND SOLVE When you step out of the shower, water on your body evaporates. Evaporation requires heat energy (the heat of vaporization), much of which comes from heat leaving your body. You feel cold because your body is transferring its heat to evaporate the water. When you are dry, there is little heat lost due to evaporation.

REFLECT This problem also explains why one feels cooler in a dry climate than in a humid climate. Your sweat evaporates more rapidly in a dry climate, taking away more heat, than in a humid climate.

3: Cold water versus cold air

Would you prefer to spend 10 minutes in a 40°F (4°C) room or in a 40°F pool? Why?

Solution

SET UP AND SOLVE Both the room and the pool are at the same temperature, but the 40°F room would be much more comfortable. This is because the specific heat of air is much less than the specific heat of water, (i.e., air will carry away less heat from your body than the water will in any time interval). Since the air carries away less heat, you are more comfortable in the room.

REFLECT Specific heat is the amount of heat needed to change the temperature of a material per unit mass and per unit temperature. Larger specific heats mean that more heat is carried away from an object.

Problems

1: Volume of a copper cup

A copper cup is filled to the brim with ethanol at 0°C. When the cup is heated to 35°C, 4.7 cm^3 of ethanol spills from the cup. What is the initial volume of the cup?

Solution

SET UP Both the cup and the ethanol expand as the temperature rises, and the difference in their expansion is equal to the volume of the spilled ethanol. We'll apply the volume expansion equation to both the cup and the ethanol, setting their difference equal to the volume of the spill. The coefficient of volume expansion is $5.1 \times 10^{-5}/\text{K}$ for copper and $75 \times 10^{-5}/\text{K}$ for ethanol.

SOLVE The change in volume of a material due to temperature is

$$\Delta V = \beta V_0 \Delta T.$$

We are given the volume of the spill, which is the change in volume of the ethanol minus the change in volume of the cup:

$$V_{\text{spill}} = \Delta V_{\text{ethanol}} - \Delta V_{\text{cup}}.$$

The initial volumes of the cup and the ethanol are the same. We'll call their common volume V_0. The temperature of both materials is 35°C. Replacing the changes in volumes yields

$$V_{\text{spill}} = \beta_{\text{ethanol}} V_0 \Delta T - \beta_{\text{copper}} V_0 \Delta T.$$

Solving for V_0, We obtain

$$V_0 = \frac{V_{\text{spill}}}{(\beta_{\text{ethanol}} - \beta_{\text{copper}})\Delta T} = \frac{(4.7 \text{ cm}^3)}{((75 \times 10^{-5}/\text{K}) - (5.1 \times 10^{-5}/\text{K}))(35°\text{C})} = 190 \text{ cm}^3.$$

The original volume of the cup is 190 cm^3.

REFLECT This is a straightforward application of volume thermal expansion. We did need to carefully note that both the copper and ethanol expanded, so the spillage was the difference of the changes in volume. We could almost ignore the change in volume of the copper, since the coefficient of thermal expansion is much smaller for copper than ethanol.

Practice Problem: What is the final volume of the cup? *Answer:* 190.3 cm^3.

2: Cooling hot tea

You wish to chill your freshly brewed tea with the minimum amount of ice, to avoid watering it down too much. What is the minimum amount of ice you should add to 2.0 kg of freshly brewed tea at 95°C to cool it to 5.0°C? The ice is initially at a temperature of -5.0°C.

Solution

SET UP The ice gains heat from the cooling water; the algebraic sum of the heats is zero. The amount of heat lost by the tea is given by the specific heat capacity equation, since the tea doesn't go through a phase change. The ice melts, so we need to include the latent heat of fusion, plus the changes due to the ice warming to 0°C and the melted ice warming to 5.0°C in calculating its heat gain of the ice.

SOLVE The heat transfer from the hot tea as it cools to 5.0°C is negative:

$$Q_{tea} = m_{tea}c_{water}\Delta T_{tea} = (2.0\text{ kg})(4190\text{ J/kg/K})(5.0°C-95°C) = -754{,}000\text{ J}.$$

Here, we used the heat capacity of water (4190 J/kg/K) for the tea. The ice must warm to 0°C, then melt, and then heat to 5.0°C. We find the heat for each segment of the ice warming. For the ice to heat to 0°C, we use the specific heat of ice (2010 J/kg/K):

$$Q_{ice} = m_{ice}c_{ice}\Delta T_{ice} = m_{ice}(2010\text{ J/kg/K})(0.0°C - -5.0°C) = m_{ice}(10{,}000\text{ J/kg}).$$

The heat needed to melt the ice is the heat of fusion for ice:

$$Q_{melt} = m_{ice}L_f = m_{ice}(3.34 \times 10^5\text{ J/kg}).$$

The melted ice must warm to the final temperature (5.0°C):

$$Q_{melted\ ice} = m_{ice}c_{water}\Delta T_{melted\ ice} = m_{ice}(4190\text{ J/kg/K})(5.0°C-0.0°C) = m_{ice}(21{,}000\text{ J/kg}).$$

The sum of these four quantities must be zero:

$$Q_{tea} + Q_{ice} + Q_{melt} + Q_{melted\ ice} = -754{,}000\text{ J} + m_{ice}(10{,}000\text{ J/kg}) +$$
$$m_{ice}(334{,}000\text{ J/kg}) + m_{ice}(21{,}000\text{ J/kg}) = 0.$$

Thus,

$$m_{ice} = \frac{754{,}000\text{ J}}{(10{,}000\text{ J/kg}) + (334{,}000\text{ J/kg}) + (21{,}000\text{ J/kg})} = 2.1\text{ kg}.$$

It takes a minimum of 2.1 kg of ice to cool the tea down.

REFLECT Despite your best effort, the tea will be watery. Putting the ice in a bag will prevent the melted ice water from mixing with the tea. Most importantly, we see how we must proceed stepwise through calorimeter problems.

3: Heat flow through three bars

A composite rod is made up of three equal lengths and cross sections of aluminum, brass, and copper. The free aluminum end is maintained at 100°C, the free end of the copper rod is maintained at 0°C. If the surface of the rod is insulated to prevent radial heat flow, find the temperature at each junction.

Solution

SET UP A sketch of the rod is shown in Figure 14.2. The heat current is the same through each segment of the rod, so we'll write the heat equations for each segment and set them equal to each other to solve the problem.

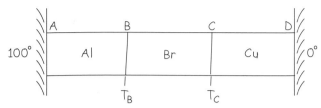

▲ **Figure 14.2**

SOLVE The heat current through any rod is

$$H = kA \frac{T_H - T_C}{L}.$$

We can write the heat current per area times length through the individual rods as

$$\frac{H_{AB}L}{A} = k_{Al}(T_H - T_C) = (205 \text{ W/m} \cdot \text{K})(100°C - T_B),$$

$$\frac{H_{BC}L}{A} = k_{Br}(T_H - T_C) = (109 \text{ W/m} \cdot \text{K})(T_B - T_C),$$

$$\frac{H_{CD}L}{A} = k_{Co}(T_H - T_C) = (385 \text{ W/m} \cdot \text{K})(T_C - 0°C).$$

These expressions must be equal to each other, since the heat currents, areas, and lengths, of the rod segments are the same. Setting the last two equations equal to each other results in

$$(109 \text{ W/m} \cdot \text{K})T_B - (109 \text{ W/m} \cdot \text{K})T_C = (385 \text{ W/m} \cdot \text{K})T_C,$$

or

$$T_B = \frac{385 + 109}{109}T_C = 4.53T_C.$$

Setting the first and last equations equal to each other gives

$$(205 \text{ W/m} \cdot \text{K})(100°C) - (205 \text{ W/m} \cdot \text{K})T_B = (385 \text{ W/m} \cdot \text{K})T_C.$$

Replacing T_B yields

$$205(100°C) - 205(4.53T_C) = 385 \ T_C,$$
$$T_C = 15.6°C.$$

Then

$$T_B = 70.7°C.$$

The aluminum–brass junction is at 70.7°C and the brass–copper junction is at 15.6°C.

REFLECT While one might expect the three equal segments to have equal temperature differences, we see that the varying thermal conductivities of the segments caused a non-uniform temperature distribution. The segment with the highest thermal conductivity (copper) had the smallest temperature difference between its ends, and the segment with the lowest thermal conductivity (brass) had the greatest temperature difference between its ends.

4: Time to melt a block of ice

A long steel rod that is insulated to prevent heat loss along its sides is in perfect thermal contact with a large container of boiling water at one end and a 3.0 kg block of ice at the other. The steel rod is 1.2 m long, with a cross-sectional of area 3.50 cm². How long does it take for the block of ice to melt if it is initially at 0°C?

Solution

SET UP We'll combine our knowledge of heat conduction with heat of fusion to solve this problem. A sketch of the problem is shown in Figure 14.3. We begin by determining the heat needed to melt the ice. We then find the rate of heat flow into the ice. With that information, we can find the time it takes for the ice to melt.

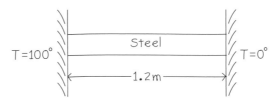

▲ **Figure 14.3**

SOLVE The heat needed to melt the ice is the heat of fusion for ice:

$$Q_{melt} = m_{ice}L_f = (3.0 \text{ kg})(3.34 \times 10^5 \text{ J/kg}) = 1.0 \times 10^6 \text{ J}.$$

The rate of heat flow is given by

$$H = \frac{\Delta Q}{\Delta t} = kA\frac{T_H - T_C}{L},$$

where k is the thermal conductivity, A and L are the area and length of the bar, respectively, and T_H and T_C are respectively the temperatures of the hot and cold sides of the bar. We find that

$$H = \frac{\Delta Q}{\Delta t} = kA\frac{T_H - T_C}{L} = (50.2 \text{ W}/(\text{m} \cdot \text{K}))(6.5 \times 10^{-4} \text{ m}^2)\frac{(100°\text{C}) - (0°\text{C})}{(1.2 \text{ m})} = 2.7 \text{ W},$$

where we used 50.2 W/m/K as the thermal conductivity of steel. The time to melt the ice is

$$\Delta t = \frac{Q_{melt}}{H} = \frac{(10^6 \text{ J})}{(2.7 \text{ W})} = 370,000 \text{ s}.$$

The time to melt the ice is 370,000 s, or 103 hours.

REFLECT We see that the thin steel bar is a relatively poor conductor of heat. Replacing the steel bar with a copper bar would increase the rate of melting by almost a factor of eight, due to the differences in thermal conductivity. Increasing the rod's diameter and shortening the rod would also increase the rate of melting.

THERMAL PROPERTIES OF MATTER

Summary

In this chapter, we extend our investigation into thermodynamics, viewing systems from both the macroscopic and the microscopic perspective and building links between them. We will learn about equations of state for materials and examine the ideal-gas equation as one example. This will allow us to build a model for the kinetic energy of individual molecules and predict the behavior of gases. We will define thermodynamic systems and examine the energy of these systems. This will lead to the first law of thermodynamics and thermodynamic processes. Four common thermodynamics processes will be highlighted, as will some implications of those processes for ideal gases.

Objectives

- Define the mole and Avogadro's number.
- Define equations of state and learn to apply the ideal-gas equation.
- Determine the kinetic energy of gases and apply it to individual particles.
- Learn about the origins of molar heat capacities for materials and gases.
- Learn and apply the first law of thermodynamics.
- Learn about the four common thermodynamic processes and apply them to find the changes in heat, work, and internal energy for thermodynamic systems.

Concepts and Equations

Term	Description
Mole	One mole (mol) is the amount of substance that contains the same number of elementary units as there are atoms in 0.012 kg of carbon 12. The number of molecules in a mole is Avogadro's number $N_A = 6.022 \times 10^{23}$ molecules per mole. The molar mass is the mass of 1 mole of a substance.
Equation of State	The equation of state is the relation between pressure, temperature, and volume of a certain amount of a substance.
Ideal-Gas Equation	The ideal-gas equation is the equation of state for an ideal gas that approximates the behavior of a real gas at low pressure and high temperature. The pressure p, temperature T, volume V, and number of moles, n, of the gas are related by $$pV = nRT,$$ where R is the ideal-gas constant. When pressure is given in Pa and volume is given in m³, $R = 8.3145 \text{ J}/(\text{mol} \cdot \text{K})$ in SI units.
Kinetic Theory of Gases	The total translational kinetic energy K_{tr} of all the molecules in an ideal gas is proportional to the temperature T and quantity of gas, n, in moles. Expressed as an equation, the total translational kinetic energy is $$K_{tr} = \tfrac{3}{2}nRT.$$ For a single molecule, the average translational kinetic energy is $$K_{av} = \tfrac{3}{2}kT,$$ where $k = R/N_A = 1.381 \times 10^{-23} \text{ J}/(\text{molecule} \cdot \text{K})$ is the Boltzmann constant.
Molar Heat Capacity	The amount of heat Q needed for a temperature change ΔT is $$Q = nC\Delta T,$$ where n is the number of moles of the substance and C is the molar heat capacity of the substance. The SI unit of C is $\text{J}/(\text{mol} \cdot \text{K})$. The molar heat capacity is the specific heat capacity times the molar mass of the material, or $C = Mc$.
First Law of Thermodynamics	When heat Q is added to a system while work W is performed by the system, the internal energy U changes by $$\Delta U = Q - W.$$ The internal energy of any thermodynamic system depends only on its state. The change in internal energy in any process depends only on the initial and final states.
Thermodynamic Processes	Common thermodynamic processes include the adiabatic process (in which no heat flows into or out of the system, so that $\Delta U = -W$), the isochoric process (in which the volume remains constant, so that $\Delta U = Q$), the isobaric process (in which the pressure remains constant, so that $W = p(V_2 - V_1)$), and the isothermal process (in which the temperature remains constant).

Properties of an Ideal Gas	The internal energy of an ideal gas depends only on its temperature, not its pressure or volume. The molar heat capacity at constant volume, C_V, and the molar heat capacity at constant pressure, C_p, for an ideal gas are related by
	$$C_p = C_V + R.$$
	For an adiabatic process in an ideal gas, both $TV^{\gamma-1}$ and pV^γ are constant, where $\gamma = C_p/C_V$.

Conceptual Questions

1: Don't hold your breath

Explain why scuba divers are taught not to hold their breath as they ascend to the surface from depths under the water.

Solution

SET UP AND SOLVE We know from fluid statics that pressure increases with depth in water. The ideal-gas equation states that pressure and volume are inversely proportional for a given temperature and quantity of gas. Ascending to the surface reduces the pressure, causing an increase in volume. By holding his breath, a scuba diver traps a quantity of air inside his lungs. As the pressure decreases on his ascent, his lungs expand, possibly damaging the diver's lung tissue. If the diver exhales during the ascent, the pressure cannot build to dangerous levels.

REFLECT This problem combines our knowledge of fluid statics and ideal gases and helps illustrate the relation between pressure and volume. High-altitude weather balloons also expand as they rise, so they are partially filled to prevent them from bursting as they ascend.

2: *pV* diagrams

One mole of helium gas is placed in a sealed container that undergoes an isochoric process that results in a doubling of the helium's pressure. Next, it undergoes an adiabatic process until the volume of the container is tripled. It then undergoes an isobaric expansion that results in a volume that is four times its original volume. Finally, the helium undergoes an isothermal compression that leaves the container with the same volume as after the first process. Sketch a *pV* diagram for this combined process.

Solution

SET UP AND SOLVE Figure 15.1 shows the resulting *pV* diagram. The diagram starts at point *a* with initial pressure p_0 and volume V_0. The segment with the isochoric process is at constant volume, represented by a vertical line to point *b*, where the pressure has doubled. In the segment with the adiabatic process, no heat is exchanged and the process follows the path to point *c*, where the volume has tripled. The segment with the isobaric process is at constant pressure, so it is represented by a horizontal line to point *d*, where the volume has increased by V_0 from point *c*. The final process is isothermic, so the segment follows the path to point *e*, where the pressure has increased to $2p_0$.

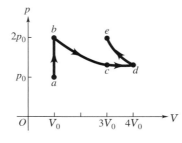

▲ **Figure 15.1**

REFLECT This problem helps clarify the differences in the four common thermodynamic problems. pV diagrams are a valuable aid in solving these kinds of problems. The diagrams will help lead us through the problem as well as provide a check on our results.

3: Internal energy in a thermodynamic process

A container of argon gas undergoes a multistep process. First, it undergoes an isobaric expansion that triples its volume. Next, it goes through an isochoric process that results in a doubling of the argon's pressure. Then it cools adiabatically by 50 K. Next, it undergoes a second isochoric process that doubles its volume. Finally, it undergoes isobaric compression that leaves it at its initial temperature. Find the total change in internal energy.

Solution

SET UP AND SOLVE The change in internal energy for an ideal gas depends only upon the temperature. Argon is an ideal gas. Since the final temperature is the same as the initial temperature, the total change in internal energy is zero.

REFLECT This complicated scenario shows that we need to focus on the important parts of the process to interpret the results properly. While it is trivial to find the change in internal energy, it would be cumbersome to find the heat added to the system.

Problems

1: Changing volume in a diving bell

A diving bell (a circular cylinder 3.0 m high, open at the bottom) is lowered into a lake. By how much does the water rise as the diving bell is lowered 75 m? The surface temperature of the lake is 25°C and the temperature at the 75 m depth is 15°C.

Solution

SET UP Figure 15.2 shows a diagram of the situation. We assume that the gas is ideal, so we can then use the ideal-gas equation to relate the surface values of pressure, temperature, and volume to the values at depth. We'll use fluid statics to relate the pressure at the surface to the pressure at depth. These two relations will be combined to find the final height of water in the bell.

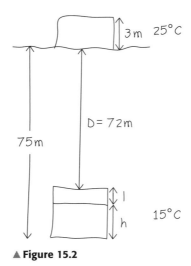

▲ **Figure 15.2**

SOLVE The same amount of gas is inside the bell both at the surface and at depth; therefore,

$$\frac{p_S V_S}{T_S} = \frac{p_D V_D}{T_D} = \text{constant},$$

where the subscript S represents the surface and the subscript D represents depth. Substituting the known values gives

$$\frac{(1.01 \times 10^5 \text{ Pa})(A)(3.0 \text{ m})}{(298 \text{K})} = \frac{p_D A l}{(288 \text{K})},$$

where we replaced the volume of the cylinder with its area times its height. There two unknowns in this equation:

$$p_D l = (288 \text{K}) \frac{(1.01 \times 10^5 \text{ Pa})(3.0 \text{ m})}{(298 \text{K})} = 2.93 \times 10^5 \text{ Pa m}.$$

We can find the pressure at depth from fluid statics, using

$$p_D = p_S + \rho g (D + l).$$

Substituting the known values yields

$$p_D = (1.01 \times 10^5 \text{ Pa}) + (1 \times 10^3 \text{ kg/m}^3)(9.8 \text{ m/s}^2)(72 \text{ m} + l),$$
$$p_D = (8.07 \times 10^5 \text{ Pa}) + (7.06 \times 10^5 \text{ Pa/m})l.$$

This equation also has two unknowns. Combining the two equations to eliminate p_D results in

$$p_D = \frac{(2.93 \times 10^5 \text{ Pa m})}{l} = (8.07 \times 10^5 \text{ Pa}) + (7.06 \times 10^5 \text{ Pa/m})l.$$

Rearranging terms gives

$$(7.06/\text{m}^2)l^2 + (8.07/\text{m})l - 2.93 = 0.$$

This is a quadratic equation with solutions $l = 0.290$ m and $l = -1.43$ m. The negative root is non-physical, so the correct l is 0.29 m. The water rises 3.0 m − 0.29 m, or 2.7 m, as the bell descends.

REFLECT This problem illustrates how the increased pressure at depth reduces the volume of gas in the diving bell. If you consider the reverse process, you can see how the volume would increase as the bell rises to the surface, as we've discussed in the questions. You can also try both examples, using a bucket of water. Submerge an inverted glass in a bucket of water, and see how the water level in the glass rises as the glass is lowered. Then use a hose to add air to the bottom of an inverted glass at the bottom of the bucket. As you raise the glass, you should see air leaving it.

2: Adiabatic compression of helium

Helium gas is expanded adiabatically from a 12 liter volume at STP to a 33 liter volume. Find the final temperature and pressure of the gas.

Solution

SET UP Since the helium is expanded adiabatically, both pV^γ and $TV^{\gamma-1}$ are constant during the process. For helium, $\gamma = 1.67$. (See Table 15.4.) Standard temperature and pressure (STP) refers to a temperature of 273 K and a pressure of 1 atmosphere.

SOLVE To find the final pressure, we use the relation

$$pV^\gamma = \text{constant} = p_1 V_1^\gamma = p_2 V_2^\gamma,$$

where the subscripts 1 and 2 refer to the initial and final points, respectively. Rearranging terms to find the final pressure, we obtain

$$p_2 = \frac{p_1 V_1^\gamma}{V_2^\gamma} = \frac{(1.01 \times 10^5 \text{ Pa})(12 \ \ell)^{1.67}}{(33 \ \ell)^{1.67}} = 1.86 \times 10^4 \text{ Pa}.$$

To find the final temperature, we use the relation

$$TV^{\gamma-1} = \text{constant} = T_1 V_1^{\gamma-1} = T_2 V_2^{\gamma-1}.$$

Rearranging terms to find the final temperature gives

$$T_2 = \frac{T_1 V_1^{\gamma-1}}{V_2^{\gamma-1}} = \frac{(273 \text{ K})(12 \ \ell)^{0.67}}{(33 \ \ell)^{0.67}} = 139 \text{ K}.$$

The final pressure is 1.86×10^4 Pa and the final temperature is 139 K.

REFLECT We see that both the temperature and pressure decreased in this adiabatic expansion. This is sensible, since there was no heat transferred into or out of the system. Thus, having a larger volume required a lower pressure and temperature.

   Would you find the same final temperature using the ideal-gas equation? If you check, you'll find that you do.

3: Isochoric and isobaric process with helium

Two moles of helium gas are taken from point *a* to point *c* in the diagram shown in Figure 15.3. Find the change in internal energy along path *abc*.

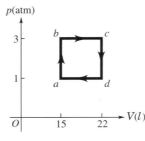

▲ **Figure 15.3**

Solution

SET UP We break the process up into two segments, one from *a* to *b* and one from *b* to *c*. The first segment is an isochoric process (constant volume) and the second is isobaric (constant pressure). We'll use the relations for those segments to determine the work, heat, and temperature changes. We'll combine the work and heat changes to find the change in internal energy.

SOLVE We find the change in internal energy along *ab* by first finding the change in temperature along *ab:*

$$\Delta T_{ab} = \frac{(p_b - p_a)V_a}{nR} = \frac{(3.03 \times 10^5 \text{ Pa} - 1.01 \times 10^5 \text{ Pa})(15 \ \ell)(10^{-3} \text{ m}^3/\ell)}{(2 \text{ mol})(8.31 \text{ J/mol/K})} = 182 \text{ K}.$$

The heat transferred during *ab* is then

$$Q_{ab} = nC_V\Delta T_{ab} = (2 \text{ mol})(12.47 \text{ J/mol/K})(182 \text{ K}) = 4500 \text{ J},$$

where we used the molar heat capacity at constant volume for helium ($C_V = 12.47 \text{ J/mol/K}$, from Table 15.4). The work during segment *ab* is zero, since this part of the process is isochoric. The change in internal energy for segment *ab* is then

$$\Delta U_{ab} = Q_{ab} - W_{ab} = 4500 \text{ J} - 0 = 4500 \text{ J}.$$

For segment *bc,* we follow the same procedure. The change in temperature along *bc* is

$$\Delta T_{bc} = \frac{p_b(V_c - V_b)}{nR} = \frac{(3.03 \times 10^5 \text{ Pa})(2.2 \times 10^{-2} \text{ m}^3 - 1.5 \times 10^{-2} \text{ m}^3)}{(2 \text{ mol})(8.31 \text{ J/mol/K})} = 127 \text{ K}.$$

The heat transferred during *bc* is

$$Q_{bc} = nC_P\Delta T_{bc} = (2 \text{ mol})(20.78 \text{ J/mol/K})(127 \text{ K}) = 5300 \text{ J},$$

where we used the molar heat capacity at constant pressure for helium ($C_P = 20.78 \text{ J/mol/K}$, from Table 15.4). The work during segment *bc* is

$$W_{bc} = p_b\Delta V_{bc} = (3.03 \times 10^5 \text{ Pa})((2.2 \times 10^{-3} \text{ m}^3) - (1.5 \times 10^{-3} \text{ m}^3)) = 2100 \text{ J}.$$

The change in internal energy during segment *bc* is

$$\Delta U_{bc} = Q_{bc} - W_{bc} = 5300 \text{ J} - 2100 \text{ J} = 3200 \text{ J}.$$

The total change in internal energy is

$$\Delta U = \Delta U_{ab} + \Delta U_{bc} = 4500 \text{ J} + 3200 \text{ J} = 7700 \text{ J}.$$

The total change in internal energy in the system is 7700 J.

REFLECT We see that, by breaking up a process into segments, determining the type of process for each segment, and knowing which variables change and which remain constant during each segment, one can easily find the change in internal energy.

How should the change in internal energy along path *adc* compare with the change along path *abc*? They should be the same for helium, an ideal gas.

Practice Problem: Find the change in internal energy along segments *ad* and *dc*, and compare their sum with the change in internal energy along path *abc*. *Answer:* $\Delta U_{ad} = 1100$ J, $\Delta U_{dc} = 6600$ J, $\Delta U_{adc} = 7700$ J.

4: Expansion process for argon

One mole of argon is initially at 25°C and occupies a volume of 35 liters. The argon is first expanded at constant pressure until its volume is doubled and then expanded adiabatically until the temperature returns to 25°C. Find the total change in internal energy, the total work done by the argon, and the final volume and pressure of the argon.

Solution

SET UP Figure 15.4 shows the *pV* diagram for the process, which we'll break up into two segments, one from *a* to *b* and one from *b* to *c*. The first segment is an isobaric process (constant pressure) and the second is adiabatic (no heat exchanged). We'll use the equations for those segments to determine the work, heat, and temperature changes, which we'll then combine to solve the problem.

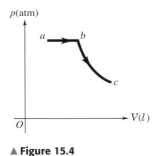

▲ **Figure 15.4**

SOLVE The total change in internal energy is zero, since the internal energy for an ideal gas depends only on temperature and the final temperature is equal to the initial temperature.

First, we find the temperature at point *b*. Segment *ab* is at constant pressure, so

$$\frac{V_a}{T_a} = \frac{V_b}{T_b}.$$

The temperature at *b* is

$$T_b = \frac{V_b}{V_a} T_a = \frac{(70 \; \ell)}{(35 \; \ell)}(273 + 25) \text{K} = 596 \text{ K}.$$

The heat supplied during *ab* is

$$Q_{ab} = nC_P \Delta T_{ab} = (1 \text{ mol})(20.78 \text{ J/mol/K})(596 \text{ K} - 298 \text{ K}) = 6190 \text{ J},$$

where we used the molar heat capacity at constant pressure for argon ($C_P = 20.78$ J/mol/K, from Table 15.4). The heat transferred during *bc* is zero, since segment *bc* is adiabatic. The total heat supplied in the complete process is 6190 J. Since the total internal energy change is zero, the heat supplied must be equal to the work done by the argon. Thus, the work done by the argon is 6190 J.

Next, we find the final volume and pressure. To find the final volume in the adiabatic process (*bc*), we use the relation

$$TV^{\gamma-1} = \text{constant} = T_b V_b^{\gamma-1} = T_c V_c^{\gamma-1}.$$

Now, $\gamma = 1.67$ for argon, so, rearranging terms to find the final volume gives

$$V_c = \sqrt[\gamma-1]{\frac{T_b V_b^{\gamma-1}}{T_c}} = \sqrt[0.67]{\frac{(596\text{ K})(70\ \ell)^{0.67}}{(298\text{ K})}} = 197\ \ell.$$

We can use the equation of state for an ideal gas to find the final pressure:

$$p_c = \frac{nRT_c}{V_c} = \frac{(1\text{ mol})(8.31\text{ J/mol/K})(298\text{ K})}{(197 \times 10^{-3}\text{ m}^3)} = 1.26 \times 10^4\text{ Pa}.$$

The final pressure is 12,600 Pa and the final volume is 197 liters.

REFLECT We again see that we need to break the process up into segments and work through each segment to find the final state variables. We also see that the *pV* graph is useful in solving problems involving thermodynamic processes.

THE SECOND LAW OF THERMODYNAMICS

Summary

In this chapter, we will complete our investigation of thermodynamics, examining thermodynamic processes and the second law of thermodynamics. Heat engines and refrigerators transform heat into work or energy in cyclic processes. Thermal efficiency and performance coefficients for engines and refrigerators will be defined. The second law of thermodynamics limits the efficiency of engines and has profound implications in many physical processes. The second law can be quantified in terms of entropy, a measure of disorder. We will examine several common cyclic processes to aid our understanding of thermodynamics.

Objectives

- Define and be able to identify reversible processes.
- Understand and analyze heat engines and refrigeration cycles.
- Learn and apply the second law of thermodynamics.
- Learn about and calculate entropy for a variety of systems.
- Gain experience from a variety of engine and refrigeration cycles.

Concepts and Equations

Term	Description
Directions of Thermodynamic Processes	Heat flows spontaneously from hotter objects to cooler objects in thermodynamic processes. A reversible or equilibrium process is a process that can be reversed by infinitesimal changes its conditions.
Heat Engine	A heat engine takes heat Q_H from a source, converts part of the heat to work W, and discards the remaining heat $\|Q_C\|$ at a lower temperature. The heat engine's thermal efficiency is $$e = \frac{W}{Q_H} = 1 + \frac{Q_C}{Q_H} = 1 - \frac{\|Q_C\|}{\|Q_H\|}.$$
Refrigerator	A refrigerator takes heat Q_C from a cold source, performs work W, and discards the heat $\|Q_H\|$ to a warmer source. The performance coefficient K is $$K = \frac{Q_C}{W} = \frac{\|Q_C\|}{\|Q_H\| - \|Q_C\|}.$$
The Second Law of Thermodynamics	The second law of thermodynamics states that it is impossible for any cyclic system to convert heat completely into work. It also states that no cyclic process can transfer heat from a cold place to a hot place without an input of work.
Carnot Engine	The Carnot engine is the most efficient heat engine. The Carnot cycle combines the reversible adiabatic and isothermal expansion and contraction between two heat reservoirs at temperatures T_H and T_C. The efficiency of the Carnot cycle is $$e_{\text{Carnot}} = 1 - \frac{T_C}{T_H} = \frac{T_H - T_C}{T_H}.$$
Entropy	Entropy is a quantitative measure of the disorder of a system. The entropy change in a reversible thermodynamic system is $$\Delta S = \frac{Q}{T}.$$ The second law of thermodynamics can be stated as "the entropy of an isolated system may increase but not decrease. The total entropy of a system interacting with its surroundings may never decrease."
Kelvin Scale	The Kelvin temperature scale is based on the Carnot cycle and is independent of the material. The zero point of the Kelvin scale is absolute zero.

Conceptual Questions

1: Cleaning your room

Your parents are always nagging you about cleaning your room. After learning about the second law of thermodynamics, you explain to your parents that it is impossible to clean your room, since it would reduce the entropy inside your room. Your mother recalls her college physics course and convinces you that you can clean your room without violating the second law. How does she convince you?

Solution

SET UP AND SOLVE She agrees with you that the entropy of a closed system can never decrease. But she notes that when you clean your room, the system consists of you plus your belongings. You can decrease the entropy of your belongings in your room by increasing the entropy of your body, as long as the total entropy increases. Therefore, you can certainly clean your room.

REFLECT This problem shows how the entropy of isolated components in a system may decrease as long as the system's total entropy increases. It also shows that you shouldn't argue with your mother, although you may want to try the argument on your father, who doesn't remember his physics course.

2: Leaving a refrigerator door open to cool a room

When the air-conditioning system at your house fails, your younger brother suggests leaving the refrigerator door open to cool the house. Is this method effective?

Solution

SET UP AND SOLVE A refrigerator cools its contents by taking heat away from them, performing work, and expelling heat to a warmer region. The expelled heat is always greater than the heat removed from the contents. The refrigerator must add net heat to its surroundings. Opening the refrigerator will result in a warmer room, so it is not an effective method of cooling the room.

REFLECT Can opening the refrigerator warm the house on a cold day? Yes, since it must expel heat to operate. It wouldn't be a very efficient heat source, but it would heat the room.

3: Water as a fuel

Some people have suggested using water as a clean fuel. The idea is to break apart water molecules into hydrogen and oxygen. When the hydrogen is burned (combined with water), it produces energy without pollution. How does the second law of thermodynamics relate to this idea?

Solution

SET UP AND SOLVE Since the breaking apart of the water and the burning of hydrogen is a reversible cycle, the net entropy must increase. The process may actually create *more* pollution, since it takes more energy to dissociate the water than is recovered by burning the hydrogen. For example, if gasoline is used to generate the hydrogen, it would take more gasoline to generate an equivalent amount of hydrogen-based power than if the gasoline were used to operate the vehicle directly.

REFLECT There is the possibility that pollution would be reduced. If the hydrogen generating plants installed high-quality pollution filters, then there may be less pollution generated overall by the plant compared with the pollution generated by many cars. However, a hydrogen-burning car will always require more total energy to operate. Hydrogen should probably be considered an alternative energy storage method.

Problems

1: Work in a heat engine

A heat engine carries 0.2 mol of argon through the cyclic process shown in Figure 16.1. Process *ab* is isochoric, process *bc* is adiabatic, and process *ca* is isobaric at a pressure of 2.0 atm. Find the net work done by the gas in the complete cycle. The temperatures at the endpoints of the process are: T_a = 290 K, T_b = 650 K, and T_c = 440 K.

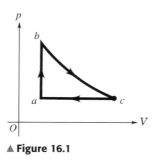

▲ **Figure 16.1**

Solution

SET UP We'll break the cycle up into processes and find the work done in each process. We'll need to find the pressure and volume at the three points before finding the work. Argon is an ideal gas, so we'll use the ideal-gas relations.

SOLVE Starting at point *a*, we find the volume V_a from the ideal-gas equation:

$$V_a = \frac{nRT_a}{p_a} = \frac{(0.2 \text{ mol})(8.31 \text{ J/mol/K})(290 \text{ K})}{(2.02 \times 10^5 \text{ Pa})} = 2.39 \times 10^{-3} \text{ m}^3.$$

At point *b*, the volume is the same as at point *a*. We find the pressure at *b*:

$$p_b = \frac{nRT_b}{V_b} = \frac{(0.2 \text{ mol})(8.31 \text{ J/mol/K})(650 \text{ K})}{(2.39 \times 10^{-3} \text{ m}^3)} = 4.52 \times 10^5 \text{ Pa}.$$

We find the volume V_c at *c*:

$$V_c = \frac{nRT_c}{p_c} = \frac{(0.2 \text{ mol})(8.31 \text{ J/mol/K})(440 \text{ K})}{(2.02 \times 10^5 \text{ Pa})} = 3.62 \times 10^{-3} \text{ m}^3.$$

With these values, we can find the work in each process. For the process *ab,* there is no work done, since it is a isochoric process. Process *bc* is adiabatic and no heat is exchanged. The work is opposite the change in internal energy:

$$W_{bc} = -\Delta U_{bc} = -nC_V(T_c - T_b) = -(0.2 \text{ mol})(12.47 \text{ J/mol/K})(440 \text{ K} - 650 \text{ K}) = 524 \text{ J}.$$

Here, we used the molar heat capacity at constant volume for argon (C_V = 12.47 J/mol/K, from Table 15.4). Process *ca* is isobaric and the work is

$$W_{ca} = p_c \Delta V_{ca} = (2.02 \times 10^5 \text{ Pa})((2.39 \times 10^{-3} \text{ m}^3) - (3.62 \times 10^{-3} \text{ m}^3)) = -248 \text{ J}.$$

Note that the work done by the gas is negative, since it is compressed in *ca*. The work for the complete cycle is the sum of the work for the three processes:

$$W = W_{ab} + W_{bc} + W_{ca} = 0 + 524\,\text{J} - 248\,\text{J} = 276\,\text{J}.$$

The gas does 276 J of work in one cycle.

REFLECT We see that, since the area inside the *pV* cycle diagram is equal to the work, we get a positive result, in agreement with the area shown in the diagram. Much of this problem is based on knowledge we acquired in Chapter 15. We are now combining the processes of that chapter into a complete cycle.

2: Efficiency of a heat engine

Find the thermal efficiency of an engine that operates in accordance with the cycle shown in Figure 16.2: (a) quantity of 2 moles of helium stored at 2.0 atm in a 10 liter vessel starts at point *a*, (b) undergoes an isochoric process to quadruple its pressure at point *b*, (c) triples in an volume in isobaric expansion to point *c*, (d) reduces to one-quarter the pressure through an isochoric process at point *d*, and (e) goes through isobaric compression, reducing its volume by one third to return to point *a*.

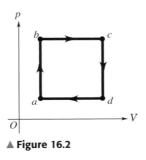

▲ **Figure 16.2**

Solution

SET UP We need the work and heat of the cycle to find the efficiency of the engine. We break the cycle up into processes and use our knowledge of isochoric (constant-volume) and isobaric (constant-pressure) processes. We are given the changes in pressure and volume, so we can proceed immediately to calculating the work in the four processes. Helium is an ideal gas, so we can use the ideal-gas relations.

SOLVE We start by determining the pressure and volume at the points *b*, *c*, and *d*. The pressure at points *b* and *c* is 8 atm $(4P_a)$ and the pressure at *d* is 2 atm (P_a). The volume at *b* is 10 liters (V_a) and the volume at *c* and *d* is 30 liters $(3V_a)$. Next, we find the work for each process. No work is done during processes *ab* and *cd*, since they are isochoric. Process *bc* is isobaric and the work is

$$W_{bc} = p_b \Delta V_{bc} = (8.08 \times 10^5\,\text{Pa})((30 \times 10^{-3}\,\text{m}^3) - (10 \times 10^{-3}\,\text{m}^3)) = 16{,}200\,\text{J}.$$

Process *da* is also isobaric and the work is

$$W_{da} = p_d \Delta V_{da} = (2.02 \times 10^5\,\text{Pa})((10 \times 10^{-3}\,\text{m}^3) - (30 \times 10^{-3}\,\text{m}^3)) = -4040\,\text{J}.$$

The total work is the sum of the work for the four processes:

$$W = W_{ab} + W_{bc} + W_{cd} + W_{da} = 0 + 16{,}200\,\text{J} + 0 - 4040\,\text{J} = 12{,}100\,\text{J}.$$

Next, we need the heat flowing into the engine. Heat flows into the engine during processes *ab* and *bc*. To find the heat flow into the engine, we need the temperature at points *a, b,* and *c.* The ideal-gas equation gives

$$T_a = \frac{p_a V_a}{nR} = \frac{(2.02 \times 10^5 \text{ Pa})(10 \times 10^{-3} \text{ m}^3)}{(2 \text{ mol})(8.31 \text{ J/mol/K})} = 122 \text{ K},$$

$$T_b = \frac{p_b V_b}{nR} = \frac{(8.08 \times 10^5 \text{ Pa})(10 \times 10^{-3} \text{ m}^3)}{(2 \text{ mol})(8.31 \text{ J/mol/K})} = 486 \text{ K},$$

$$T_c = \frac{p_c V_c}{nR} = \frac{(8.08 \times 10^5 \text{ Pa})(30 \times 10^{-3} \text{ m}^3)}{(2 \text{ mol})(8.31 \text{ J/mol/K})} = 1460 \text{ K}.$$

The heat flow in process *ab* (constant volume) is

$$Q_{ab} = nC_V \Delta T_{ab} = (2 \text{ mol})(12.47 \text{ J/mol/K})(486 \text{ K} - 122 \text{ K}) = 9080 \text{ J}.$$

where we used the molar heat capacity at constant volume for helium ($C_V = 12.47 \text{ J/mol/K}$, from Table 15.4). The heat flow in process *bc* (constant pressure) is

$$Q_{bc} = nC_P \Delta T_{bc} = (2 \text{ mol})(20.78 \text{ J/mol/K})(1460 \text{ K} - 486 \text{ K}) = 40,500 \text{ J},$$

where we used the molar heat capacity at constant pressure for helium ($C_P = 20.78 \text{ J/mol/K}$ from Table 15.4). The total heat flowing into the engine in one cycle is therefore

$$Q_H = Q_{ab} + Q_{bc} = 40,500 \text{ J} + 9080 \text{ J} = 49,600 \text{ J}.$$

The efficiency of the engine is

$$e = \frac{W}{Q_H} = \frac{12,100 \text{ J}}{49,600 \text{ J}} = 24.4\%.$$

REFLECT We see that the engine is 24.4% efficient. To find the efficiency, we started by determining the state variables for the points on the *pV* diagram. Then we found the work and heat flow in the cycle and combined this information to solve the problem.

3: Entropy change in melting ice

A heat reservoir at 50°C is used to melt 25 kg of ice at 0°C. What is the entropy change in the melted ice? What is the entropy change in the reservoir? What is the total entropy change in the system?

Solution

SET UP Entropy change in a reversible process is the heat transferred divided by the temperature of the material. We can find the heat required to melt the ice, which must be equal to the heat provided by the reservoir.

SOLVE The heat required to melt the ice is given by the heat-of-fusion relation and is

$$Q = mL_f = (25 \text{ kg})(335 \times 10^3 \text{ J/kg}) = 8.375 \times 10^6 \text{ J},$$

where we used the latent heat of fusion for ice ($335 \times 10^3 \text{ J/kg}$). The change in entropy for the water is

$$\Delta S_{ice} = \frac{Q}{T} = \frac{8.375 \times 10^6 \text{ J}}{273 \text{ K}} = 30,700 \text{ J/K}.$$

The change in entropy for the reservoir is equal to the heat leaving the reservoir, divided by the reservoir's temperature. The reservoir loses as much heat as the ice gains. The entropy change is

$$\Delta S_{reservoir} = \frac{-Q}{T} = \frac{-8.375 \times 10^6 \text{ J}}{323 \text{ K}} = -25{,}900 \text{ J/K}.$$

The total change in entropy is the sum of the entropies for the ice and reservoir:

$$\Delta S_{total} = \Delta S_{ice} + \Delta S_{reservoir} = 30{,}700 \text{ J/K} - 25{,}900 \text{ J/K} = 4800 \text{ J/K}.$$

The entropy of the ice increases by 30,700 J/K, the entropy of the reservoir decreases by 25,900 J/K, and the total entropy of the system increases by 4800 J/K.

REFLECT The entropy for the reservoir decreased, but this does not violate the second law, since is the system is the combination of the ice-water and the reservoir. The total change in entropy is positive, as expected.